The Geography of the Ocean

Despite the fact that the vast majority of the earth's surface is made up of oceans, there has been surprisingly little work by geographers which critically examines the ocean-space and our knowledge and perceptions of it. This book employs a broad conceptual and methodological framework to analyse specific events that have contributed to the production of geographical knowledge about the ocean. These include, but are not limited to, Christopher Columbus' first transatlantic journey, the mapping of nonexistent islands, the establishment of transoceanic trade routes, the discovery of large-scale water movements, the HMS Challenger expedition, the search for the elusive Terra Australis Incognita, the formulation of the theory of continental drift and the mapping of the seabed.

Using a combination of original, empirical (archival, material and cartographic), and theoretical sources, this book uniquely brings together fascinating narratives throughout history to produce a representation and mapping of geographical oceanic knowledge. It questions how we know what we know about the oceans and how this knowledge is represented and mapped. The book then uses this representation and mapping as a way to coherently trace the evolution of oceanic spatial awareness.

In recent years, particularly in historical geography, discovering and knowing the ocean-space has been a completely separate enterprise from discovering and colonising the lands beyond it. There has been such focus on studying colonised lands, yet the oceans between them have been neglected. This book gives the geographical ocean a voice to be acknowledged as a space where history, geography and indeed historical geography took place.

Anne-Flore Laloë, after graduating from the University of Exeter, was Curator of Historical Collections at the Marine Biological Association of the UK. Since January 2015, she has been Archivist at the European Molecular Biology Laboratory.

Studies in Historical Geography
Series Editor: Professor Robert Mayhew
University of Bristol, United Kingdom

Historical geography has consistently been at the cutting edge of scholarship and research in human geography for the last fifty years. The first generation of its practitioners, led by Clifford Darby, Carl Sauer, and Vidal de la Blache, presented diligent archival studies of patterns of agriculture, industry, and the region through time and space.

Drawing on this work, but transcending it in terms of theoretical scope and substantive concerns, historical geography has long since developed into a highly interdisciplinary field seeking to fuse the study of space and time. In doing so, it provides new perspectives and insights into fundamental issues across both the humanities and social sciences.

Having radically altered and expanded its conception of the theoretical underpinnings, data sources, and styles of writing through which it can practice its craft over the past twenty years, historical geography is now a pluralistic, vibrant, and interdisciplinary field of scholarship. In particular, two important trends can be discerned. First, there has been a major "cultural turn" in historical geography that has led to a concern with representation as driving historical-geographical consciousness, leading scholars to a concern with text, interpretation, and discourse rather than the more materialist concerns of their predecessors. Second, there has been a development of interdisciplinary scholarship, leading to fruitful dialogues with historians of science, art historians, and literary scholars in particular, which has revitalized the history of geographical thought as a realm of inquiry in historical geography.

Studies in historical geography aim to provide a forum for the publication of scholarly work, which encapsulates and furthers these developments. Aiming to attract an interdisciplinary and international authorship and audience, Studies in Historical Geography will publish theoretical, historiographical, and substantive contributions meshing time, space, and society.

The Geography of the Ocean

Knowing the ocean as a space

Anne-Flore Laloë

Routledge
Taylor & Francis Group

LONDON AND NEW YORK

First published 2016 by Routledge

2 Park Square, Milton Park, Abingdon, Oxfordshire OX14 4RN
711 Third Avenue, New York, NY 10017

Routledge is an imprint of the Taylor & Francis Group, an informa business

First issued in paperback 2018

British Library Cataloguing in Publication Data
A catalogue record for this book is available from the British Library

Library of Congress Cataloging in Publication Data
Names: Laloë, Anne-Flore, author.
Title: The geography of the ocean : knowing the ocean as a space /
   by Anne-Flore Laloë.
Description: Farnham, Surrey, UK ; Burlington, VT : Ashgate, 2016. |
   Series: Studies in historical geography | Includes bibliographical
   references and index.
Identifiers: LCCN 2015041448 (print) | LCCN 2015046650 (ebook) |
   ISBN 9781409421436 (hardback : alk. paper) | ISBN 9781409421443
   (ebook) | ISBN 9781472400857 (epub)
Subjects: LCSH: Ocean—History. | Ocean and civilization. |
   Historical geography. | Discoveries in geography—History.
Classification: LCC GC29 .L35 2016 (print) | LCC GC29 (ebook) |
   DDC910.9162—dc23
LC record available at http://lccn.loc.gov/2015041448

ISBN: 978-1-4094-2143-6 (hbk)
ISBN: 978-1-138-54650-9 (pbk)

Typeset in Times New Roman
by Apex CoVantage, LLC

Contents

Figures

1 Introduction

Human geographers have, overall, an ambiguous relationship with the ocean. At times, they ignore it, thereby reducing their study to the earth's land surfaces; their geographical perspective thus finds itself incomplete and ignorant of what it might be missing. In other instances, coastlines are considered and their inherent liminality explored, both as regards what is found along it and on either side of it, though the ocean itself is not fully engaged with. In these cases, the ocean helps define the land or coast by opposition, saying what it is not. Finally, when thought is given to the ocean beyond its coasts, it is usually limited to specific discourses on environmental history, the plundering of oceanic resources or complex issues regarding territorial waters and, again, resources. In these instances, the ocean is considered primarily for what it contains or potentially contains, not as a space in itself. This materialistic approach divorces the ocean as a container of resources from the ocean as a space. The ocean itself, therefore, with its physical attributes, wavering surface, tidal rhythms and unbounded connectivity, and what these characteristics mean for human and nonhuman interrelations with it, are rarely at the center of human geographers' oceanic engagement. Studies might capitalize on one of these aspects, either directly or metaphorically, but, on the whole, this means that the understanding of the geographical ocean generally falls short of engaging with the ocean as a geographical space. This book will address this issue and demonstrate that a wholly geographical study of the ocean is both possible and fruitful. Furthermore, it will show that such an approach enables geographers to engage critically with traditional geographical concerns such as mobility, governance, and cartography in novel ways that fully comprehend the specific mobilities, spatialities, and materialities of the ocean. This will reveal an ocean that can fully be a part of geographical discourse, rather than sitting at its fringes, and will crucially move toward a truly global sense of geography. This geographical ocean is what will be called the ocean-space.

The term ocean-space was defined by Philip Steinberg (1999b, 367–368) as one "that is intended to capture both the specificity of the world ocean and the fluidity between the study of landward and seaward domains, as both are socially and physically linked through linked dynamics." In previous work, I used it to put emphasis on the physically geographical space of the ocean as opposed to the word *ocean*, which, I posited, referred to as the culturally constructed oceanic space that focuses on costal zones and human interactions (Laloë 2009). In this latter sense, the notion captured by *ocean-space* moves away from an understanding of the ocean as a two-dimensional surface and instead focuses attention on the ocean as a space with three- and four-dimensional physical characteristics and puts these at the core of our geographical engagement with it. In other words, the term ocean-space assumes, rather than makes apologies for, the fact that the ocean's physical characteristics and their spatial aspects are not fixed and that these movements are inseparable from the ocean's geography. Ocean-space therefore encapsulates a physical and spatial ocean, which certainly allows us to go beyond traditional understandings of geographical sites as fixed and acted upon; instead, ocean-space focuses on the ocean's physical characteristics that both define and allow for its specific human geographies.

It thus becomes clear, also, that the ocean-space is different from the unhyphenated ocean space, which is typically used in the context of maritime governance relating to policy making or fisheries (see Pirtle 2000; Vallega 2001; Hoagland et al. 2003). Within these contexts, the ocean space can be said to in fact refer to oceanic space, which is the ocean's surface or resources upon which to implement rules and regulations. This ocean space is construed as an extension of terrestrial space, and its treatment is redolent of notions of governance and statehood. This is an important differentiation, since it brings to the fore the ocean-space's emphasis on its own physical characteristics as opposed to the ocean space's subjection to terrestrial protocols of measuring and distributing surfaces. Most significantly, my focus on ocean-space rather than ocean space reemphasizes the fact that I will not be considering the ocean as simply a stage upon which some human events take place, but rather move the ocean-space to the foreground and consider its place in shaping our knowledge about it.

As a whole, then, this book is about knowing the ocean as a geographical space, or indeed knowing the ocean-space. Throughout these pages, I will examine how knowledge about the ocean-space has been and is being produced in terms of scientific and cultural paradigms that are at the core of geography today. Through discussing how knowledge about the ocean-space was acquired and produced, this book will investigate the geography of the ocean-space from the bottom upward, considering the succession of ontological steps that come together as the body of knowledge that we have about the earth's ocean.

The methodology of this book will be threefold. Foremost, it will be driven by archival and historical material, including charts, survey documents, and sailing directions. Together, these will be taken to be representative of a particular type of knowledge about the ocean and as a specific way of knowing the earth. This follows Thrower (1981, 1) who argues that "as a branch of human endeavor, cartography has a long and interesting history that well reflects the state of cultural activity, as well as the perception of the world, in different periods." Accompanying this archival study will be a study of knowledge in its widest sense. This will allow me to engage with and challenge historical notions of knowledge and concepts of knowledge production as regards the understanding of the ocean-space. Certainly, the representation of knowledge about the earth is contingent on evolving paradigms of knowledge and representations of knowledge; archival material must therefore be examined in light of this. The third aspect of this study will be concerned with technological developments. The exploration of the ocean-space has always been and will inevitably remain heavily reliant on technology; any study of the ocean-space must therefore consider the role of technology. Specific technologies, such as sounding devices, ships, or measuring buoys will be studied in light of how they both enabled and limited exploration of the ocean-space. By considering cartographical representations alongside cultural and technological settings, this book will therefore offer a comprehensive study of the geography of the ocean-space.

The time frame for this study seeks to cover humankind's main large-scale interactions with the ocean, meaning that the primary focus will be oceans, rather than seas, though these will not be excluded. Therefore, over four chronological chapters, I will be considering, in turn, notions of ocean-space between 1492 and the beginning of the Scientific Revolution, the ocean-space and the Enlightenment, the birth of oceanographic science, and the ocean-space in the twentieth century and beyond. Each of these divisions will be justified in due course, but the emphasis here is on this book's wide-angle view. This is to reflect the ocean-space's own timeless characteristics and its unbounded nature. This approach also distinguishes this book from existing literature in the fields of both maritime history and historical geography, which have a tendency to focus on single historical events or at least much shorter timescales, such as the lifetime of a person, a ship, or a technology (e.g., Lavery 1989; Allen 2002; Collingwood 2003; Corfield 2003).

Similarly, this book does not focus on a single ocean or sea; its approach instead seeks to be relevant to all of the earth's traditional oceans. This is a logical consequence of the use of the term ocean-space, since there is indeed no natural way of dividing the ocean-space based on physical characteristics. The intellectual background to this approach is discussed in Chapter 2. Since the ocean-space is deeply concerned with the materialities and spatialities, this signifies that this book is on the whole concerned with the so-called World Ocean, rather than any specific ocean, such as the Indian Ocean or the Arctic Ocean. However, the documents examined will heavily influence the areas discussed, and this explains the Atlantic Ocean bias, which itself follows from a Eurocentric outlook. Nonetheless, these discussions should not be taken as specific but instead as representative of a more general ocean-space. This global perspective further distinguishes this book from existing literature, which has typically centered studies on oceanic basins or trade routes.

Existing literature on the subject of oceans and history is vast, but its general focus is on the human interactions on and around the world's oceans. The historical subdiscipline of maritime history, for instance, is primarily concerned with how humans explored, exploited, governed, militarized, and navigated the ocean. These histories typically portray the ocean as a site for thematic history that frames a discussion. In other words, while considering transoceanic trade, colonization, exploitation, and migrations, and thinking about how these shaped the nations and cultures situated along their coasts, maritime history does not fully engage with the way in which the ocean's physical characteristics actively shape humanity's interaction with it. The focus of maritime history is to discuss the ocean and the ocean space, not the ocean-space as defined here. As a consequence, I posit that it is limited by the ocean rather than enabled by the ocean-space, reflecting humans' own limitations when interacting with the ocean-space.

By moving beyond this anthropocentric approach and placing the ocean-space at the center of this study, the ocean-space can therefore become known for its own characteristics, rather than in spite of them. This will reveal that these materialities shaped and drove human interactions with the ocean-space and that these are richer than humans' movements on the ocean's surface or their plundering of resources. By the end of this book, the ocean-space will have emerged as a space that, while also having a history, has an established human geography.

2 Intellectual setting

The earth's oceanic space is a physical environment, constituted of interconnecting oceans that can be constructed as facts and figures. As such, the Atlantic Ocean, for example, can be numerically asserted as spreading 93.7 million square kilometers, covering 18.4 percent of the earth's entire surface, and almost 25.9 percent of the area of the World Ocean (Sverdrup et al. 2003, 50).[1] This makes it the world's second largest ocean after the Pacific Ocean, which covers 165.25 million square kilometers and covers just under half of the area of the World Ocean and a third of the entire surface of the earth. The entire volume of water held in the Atlantic Ocean is 312.2 million cubic kilometers, which is approximately 23 percent of the volume of the earth's entire oceanic waters (Sverdrup et al. 2003, 50).[2]

The World Ocean's constituent parts, notably the Pacific, Atlantic, Indian, Arctic, and Southern Oceans, each have specific physical characteristics that, to an extent, differentiate them, but have also been formally delineated by the International Hydrographic Organization (IHO), which set geographical coordinates that are then ratified by the IHO's member states as an ocean's official limits. The oceans' shapes and boundaries are, however, mobile, thanks to plate tectonics: thus the Mid-Atlantic Ridge, which separates tectonic plates—the North American plate from the Eurasian plate in the North Atlantic and the South American plate from the African plate in the South Atlantic—is active and pushing the plates apart at a speed of one to ten centimeters per year (Mayhew 1997, 330).[3] Thus, the Atlantic Ocean is expanding. Consequently, the Pacific Ocean is shrinking.

The waters of the world's oceans are moved by a number of major currents, which are part of a complex system of thermohaline circulation (Wijffels et al. 1992). These elaborately interact, carrying hot and cold waters across vast distances. Among these currents is the Gulf Stream, which warms Northern Europe and which was first mapped by Benjamin Franklin in 1785 (Franklin 1806, Figure 173). The Gulf Stream flows between 30 Sv and 150 Sv, where one Sverdrup (Sv) is equivalent to the flow of 10^6 cubic meters of water per second (Johns et al. 1995, 817).[4]

The average depth of the World Ocean is 3,682 meters, and its deepest point is in the Mariana Trench, located in the Western Pacific Ocean, and reaches a depth of 10,911 meters (Charette and Smith 2010, 113).

Such figures as these about the World Ocean construct it according to a Euclidian geometry that is concerned with objectivity and specified as immutable. Understanding space in this way perceives it as a homogenous surface upon which actions are performed. It is the "necessary representation" that "supplies the basis for external phenomena" (Kant 2003, 24). This is the ocean that is charted in official hydrographic documents and is apparently removed from cultural influences.

In parallel to this outlook, the ocean can also be read as a space that is experienced personally and which is perceptually formulated and culturally influenced. In this context, Friedrich Nietzsche and José Ortega y Gasset both argue that "there are as many different spaces as there are points of view" (Kern 2001, 132). For example, the photographs that follow are representative of the space that I have experienced and perceive of as the Atlantic Ocean. They are illustrative of *my* Atlantic Ocean, largely detached from the figures presented earlier.

These two ways of viewing the oceans depict very different ways of understanding space, respectively, representative of space viewed as either absolute or relative. Yet between these two apparently exclusive oceans lies the ocean as a geographical ocean-space, which has, to a large extent, been ignored. The term "ocean-space" is here used to put emphasis on the geographical space of the ocean, both absolute and relative, standing opposed to simply the term "ocean," which can be construed as denoting a culturally constructed space that is weighted toward the ocean-space's coastal areas where humans have interacted with it.

Figure 2.1 Looking westward from the Isles of Scilly—November 2007.

Figure 2.2 Stormy weather closing in on the Baie de Quiberon, France—Easter 2008.

It is the ocean-space, which is at the center of this book, through examining knowledge production from a historical perspective and focusing on the circulation of knowledge on, across, and around the ocean-space, and considering ways in which geographical knowledge about space is produced. By exploring, from a historical perspective, how the ocean-space is as a geographical space, this book highlights that processes of knowledge production are intrinsically linked to a variety of cultural, scientific, and technical influences, which invariably shape knowledge outputs. Following the trend that emerged in the 1980s in the history of science, this study builds on the idea that science "is a practical activity, located in the routines of everyday life" or, following Donna Haraway, that science is "situated knowledge" (Secord 2004, 657; Haraway in Secord 2004, 658). Knowledge can thus be seen as a cultural variable that has to be negotiated as such, while bearing in mind certain immutable facts. At the same time, geographical knowledge of the ocean-space is characterized by particular physical characteristics that dramatically shape both the ocean-space itself and how it can be known geographically. One example of this is the fact that human exploration of the ocean-space is heavily reliant on technology whose limitations, in turn, place boundaries on exploration. As technological advances are made, human ability to explore the ocean-space also changes. Thus different periods have perceived, been able to perceive, and produced knowledge about the ocean-space in diverging but characteristic ways.

This book adopts a broad historical outlook in order to acquire a diachronic perspective that allows us to apprehend the ocean-space both as an absolute and a relative space, but also as the geographical entity that lies in between. By taking a wide position, this thesis diverges from the "small stories" approach that is at present popular within the field of historical geography and instead investigates big stories and large-scale questions that contribute to geographical knowledge production about the ocean-space (Lorimer 2003). Whereas small-scale, in-depth contextualizations are useful and important to understand specific and perhaps key moments of history or geography, these stories can rarely be projected onto larger scales, be they regional, national, continental, or global. As Secord argues, exclusively following such a method signifies that "we end up with a rich array of research that somehow adds up to less than the sum of its parts" (Secord 2004, 660). Conversely, by choosing to focus on the narrative of scientific knowledge production over four centuries with regard to a large geographical surface area, a number of small stories are connected within a historical narrative. The historical geography of the ocean-space put forward here examines archival material through a wide-angle lens and presents a diachronic (though not comparative) understanding of the ocean-space as a geographical space that was in turn discovered, mapped and measured, sounded, and dredged. This approach allows me to fathom, from the bottom up, the process of scientific knowledge production of a specific geographical space as a single narrative while also acknowledging the importance of small stories.

As a whole, this book weaves together sociocultural, scientific, religious, and technological stances that were fundamental in creating the ocean-space as the geographical space that it is known as today. Through considering these elements as linked within the enterprise that turned the ocean from an unknown non-space to a located, geographical entity, this work is situated within the historiography of science, focusing on the places of geographical knowledge.[5]

Throughout this "intellectual setting" chapter, I consider how space can be known from a historical perspective. Examining the historical geographies of space and the way that these are understood in the present will highlight mechanisms of knowledge making from a historical perspective. This will specifically consider how geographical knowledge is historically produced and, with particular regard to information about distant places, what scientific and social mechanisms were initiated in order to safeguard information. The question of "how far [information] could be trusted" was instrumental in the establishment of networks of trust that would be central to the scientific method (Driver 2004b, 73). I further discuss issues of trust between those who collected scientific samples, such as fauna, flora, or sediment, and those who analyzed them, and question how this trust was made. This is what Withers (1999, 498) calls the "historical geography of trust" and what Mayhew (2005, 75) examines as the "construction of credibility in early modern scientific practice." Considering how human relationships affected the production of knowledge is central to understanding the twice-removed connection between the Atlantic ocean-space and the scientists studying it, especially in considering how the different locales of geography, from the ship to the field to the laboratory, have to be negotiated so as to comprehend the ocean-space as not simply a site, but a space. This, in turn, applies the debates introduced in this current chapter to the ocean-space generally and the Atlantic Ocean particularly. This will discuss the idea of the Atlantic Ocean as a historical and geographical space. Klein and Mackenthun (Dening 2004, 13) argue that "the ocean has often been read as an empty space, a cultural and historical void, constantly traversed, circumnavigated and fought over, but rarely inscribed other than symbolically by the self-proclaimed agents of civilization." I will argue against this approach from a wider perspective and highlight instances in which the Atlantic Ocean has, instead, been imbued with meaning. Instead, the idea of the Atlantic Ocean as "a European invention," both historically and geographically (Armitage 2002, 12) will be debated.

Together, the remaining chapters will view the debates raised here in relation to historical and archival documents. These chapters are organized chronologically and thematically. Chapter 3 studies the period from 1492 to the beginning of the Enlightenment, covering the period during which the Atlantic Ocean became known geographically. Chapter 4 focuses on the Enlightenment and describes measuring and geographical pinpointing of the Atlantic ocean-space. Chapter 5 considers deep ocean exploration during the nineteenth century as the depths of the Atlantic Ocean were studied. This study ends with the twentieth century.

Ultimately, this book locates the ocean-space as locales of geography. By considering the Atlantic Ocean in terms of historical geography, this book positions the ocean-space at the center of geographical understandings of space, where space is both absolute and relative. As the circulations of knowledge are highlighted, it is also situated within cultural and scientific contexts to construct a complex picture, which reconciles a factual ocean-space and a known ocean.

2.1 Knowing space historically

There are many ways in which space can be thought of and constructed historically in light of theoretical, cultural, and scientific outputs. By examining the place of networks in the collection of information and the making of scientific spaces, this section will consider geographical knowledge production about distant places.

Knowing space, whether physically, through human-environment interactions; culturally; or theoretically is, broadly, the subject of geography. Through the study of specific areas of the earth and its characteristics, facts are acquired and knowledge produced about particular features of space. In the case of the ocean, lists of facts such as the aforementioned come to constitute knowledge about the space itself. Within geography, historical geography is concerned with geographies of the past and the way in which these relate to the present. Indeed, "the geographical processes which have shaped the modern world and the ways in which the past is understood and culturally represented in the present are central concerns within historical geography" (Nash 2000, 15). Methods of historical geography include the reconstructions of past geographies and mapping their changes over time, and considering specific geographical zones in a comparative manner so as to get a historically coherent overview of an area. Through focusing on localities, telling personal stories, or examining company histories, space is studied historically in relation to specific cultural or societal elements.[6] This manner of knowing space historically, as Driver (1988, 504) notes, "is no luxury; on the contrary, it is an essential part of doing human geography" and seeks to offer a dynamic picture of the past through discourses that are invariably mediated by the present.

With regard to this book, such a historical approach to the geography of the ocean includes considering the different ways of interpreting the World Ocean through the variety of cultures whose histories are linked to it. These interpretations, however, can reflect an academic cultural context as much as they do the ocean's history itself, meaning that wider contexts are important to understand spaces. Here I will show how this is the case, considering particular historical geographies of faraway spaces and historical geographies of science. Analyzing how faraway spaces and spaces of science have been examined within the framework of historical geography will enable me to draw useful parallels and extract strands that are helpful when thinking about the ocean-space historically and geographically. Together, these themes will outline the ways that are most useful to understand the ocean-space as historical and geographical spaces. These specific methods will help construct a wider methodological framework through which to consider the historical geography of the ocean-space, and which will be applied to archival and cartographic material in the following chapters.

2.1.1 Historical geographies of space

Space can indubitably be understood in a variety of ways. From the perspective of historical geography, recent studies have focused on the mechanisms by which faraway spaces became known. For instance, Driver and Martins discussed space in terms of tropical vision and "tropicality" (Martins 1998 and 2000; Driver and Martins 2002, 2005a, 2005b, and 2006). Ogborn (2000, 2002, and 2004) has studied the ways in which distant spaces became connected in the advent of globalization and through networks of communication. Ryan (1997) has engaged with the representation of imperial space through the medium of photography. Withers and Livingstone examined the geographical element of the Enlightenment and considered the role of science in producing knowledge at a distance (Livingstone and Withers 1999a; Livingstone 2003; Withers 2007). These works are representative of a specific way of doing historical geography, which considers one particular perspective and applies it in-depth to a discrete place or a certain kind of space. Together, these various ways of thematically considering the historical geography of space offer a varied and stimulating picture of the past, while the body of literature that they build is helpful both in highlighting specific trends and in bringing forth a set of methods that enables us to think about geography in specific historic ways. This literature creates a lens through which to understand space historically and, here, to perceive the ocean-space within a wider geographical context.

However, these works are intrinsically limited both by their focused methodology and because they usually concentrate on specific geographical locales. This book seeks to go beyond thematic perspectives by instead replacing space at the center of my discussion. Considering how the ocean-space has been understood in relation to a variety of factors over four centuries will foreground that knowledge about space is the result of the complex interplay between a multitude of influences, offering a more comprehensive understanding of the processes at play in making knowledge.

This chapter examines specific historical geographies of space with an emphasis on faraway locations and especially the geographical imagination of distant, uncharted lands with a focus on the tropics and the idea of "tropicality" (Driver and Martins 2005b, 3). This leads on to a discussion of what Withers (1999, 498) calls the "historical geography of trust" and the question of the transmission of knowledge across the earth and, specifically, the question of networks of knowledge and the relationship between faraway places and what Miller (in Driver 2004b, 82) dubs the "centers of calculation." This then allows us to consider attempts made to homogenize the making of knowledge about these places and, in particular, scientific efforts to achieve homogenization through the publication of guidelines and standardization. This focuses on the historical geographies of science in the eighteenth and nineteenth centuries. Finally, I consider the space of the ship, which is central to much knowledge production about the ocean-space, in relation to the geographies of science arguing that the ship is both a "scientific instrument" and a laboratory (Sorrenson 1996, 221). These ways and methodologies of making knowledge about space will then be applied to the ocean-space in later chapters.

Geographical imaginations of tropical spaces

The knowledge that we have about places, whether we have visited them or not, is amalgamated in a multilayered manner, "constructed in a variety of ways, through experience, learning, memory, and imagination" (Driver and Martins 2005b, 3). This multifaceted way of making knowledge about space is especially central to the production of knowledge about tropical spaces, to which, between 1500 and 1900, few went but many imagined. In this respect, the tropics and the ocean-space are similar: the geographies of both spaces were reliant on little information, but many representations and imaginings seeped into cultural consciousness nonetheless. Following Driver and Martins (2005b, 3), who focus on the "ways in which tropical places are encountered and experienced, the significance of travel for the process of making knowledge about these places, and the relationship between geographical difference and generalized notions of 'tropicality,'" this section will formulate an understanding of how knowledge about the ocean-space can be constructed, presenting the literature of the historical geography of faraway, unknown places and emphasizing how their discourse might be useful later in this chapter in the analysis of the historical geography of ocean-space.

Indeed, the aim here is to question how the ocean-space was discovered, classified, and geographically constructed. From this perspective, discourses of postcolonialism and imperialism indubitably underpin European constructions of the ocean-space and are essential to mechanisms of knowledge production and circulation about the ocean-space. Indeed, as Steinberg (2009, 481) points out, "several scholars of empire have noted, the idea of an intervening, marine space between the core and the periphery was crucial to the discursive and material workings to the era's European empires." Relating to the Atlantic Ocean, the works of Fanon (1952 and 1961), Césaire (1969), Walcott (1986), and Gilroy (1993) bear witness to the importance of these discourses. However, expanding a discussion of postcolonialism and imperialism is not fully within the scope of this book and, therefore, I will not consider further Said (1994 and 2003) and Bhabha (2004), who examine "the enormously systematic discipline by which European culture was able to manage—and even produce—the orient politically, sociologically, militarily, ideologically, scientifically, and imaginatively during the post-Enlightenment period" (Said 2003, 3). These discourses, while they are of course central to the idea of tropicality, are not essential with regard to the ocean-space.

With regard to the construction of knowledge about the ocean-space, focusing on European visions of the tropics during the Age of Exploration is useful in several ways. First, as Driver (2004a, 2) writes about the tropics:

> Whether the adjective 'tropical' denotes a particular kind of experience, a look, a species, a landform, a soil or a meteorological event, the term carries with it a powerful array of associations which *may or may not be tied very specifically to a particular geographical zone or location.*

(My emphasis)

Certainly, what is important is not the geographical location but the cultural imagination of the tropics as a space. What is called "tropicality" is, here, more useful to understanding the tropics than pinpointing their geographical location. The same is true of the ocean-space whose geographical imagination is, at this stage, more important than any specific physical characteristics. Duncan (2000, 34) asserts that the geography of the tropics for nineteenth-century Europeans was "imaginary" because it was "a cultural construction of nature with tenuous links to the reality of actual places." Similarly, it can be posited that the cultural construction of the distant ocean-space speaks more than the exact location of these seas. Crucially, generalized notions about the tropics are like vague thoughts about the ocean: they function as inferred contentions of geographical discourse that allow us to say less about the specific locales themselves than as what their representation says about us.

The second reason why thinking about the tropics is helpful here is linked to the physical geography of the earth: the tropics are usually connected to Europe by the oceans, and crossing the ocean-space was the most common way of reaching them. Therefore, the first views of the tropics were from a seaborne perspective. From aboard ships, surveyors, naval officers, and midshipmen drew, sounded, and measured the tropical environment (see Driver and Martins 2002). Thus the view from the sea was a significant contribution to geographical knowledge of faraway places and was instrumental in the shaping of the idea of both the tropics and the ocean. As Arnold (2005, 141) writes:

> We are now so accustomed to arriving in new places by air that we have largely lost sight of the importance of the sea and coastal views to the travelers of an earlier age, not only in framing their physical approach to the land but also in informing their intellectual and emotional approach as well. This was particularly significant for the globe-girdling naturalists of the late eighteenth and early nineteenth centuries. New continents and unknown islands, which announced themselves first by sea, created impressions that were not then easily erased.

Because the view of the tropics from the vantage point of a ship was typically the first one that Europeans would experience, ideas of tropicality were surely shaped by this sea perspective. In fact, Arnold (2005, 141) continues:

> Accounts of the tropics commonly began with descriptions of arriving (often, with heightened effect, at dawn) at some small harbor against the dramatic backdrop of dense tropical forests and to the accompaniment of the strange sounds of tropical birds and animals.

Indeed, travelers' and naturalists' descriptions of the tropics are rife with commentaries on sea crossings, harbors, and creeks (see Darwin 1958 and von Humboldt 1995 for examples of this). Thus the influence of the marine perspectives upon the idea of tropicality highlights the tight entwinement between various geographies and emphasizes that all approaches have to be considered as a whole rather than separately.

A third justification for discussing tropics here is methodological, as parallels can be drawn between the study of the tropics during the eighteenth and nineteenth centuries and that of the ocean over the same period. Specifically, both the study of the tropics and the study of the ocean-space sought to record observations directly in the field. As Greppi (2005, 23) notes, drawings made "on the spot (. . .) convey the spirit of a mode of landscape representation in which true knowledge of the natural world—and its botanical, zoological, human, and aesthetic forms—is based on direct observation of the field." Following Latour and Woolgar (1986), the advantages of in situ observation to make scientific facts is crucial: recording observations on site sought to give credence to the phenomena being recorded, as it was allegedly neither distorted nor embellished. As they contend:

> Many aspects of science described by sociologists depend on the routinely occurring minutiae of scientific activity [in the laboratory]. Historic events, breakthroughs and competition are examples of phenomena which occur over and above a continual stream of ongoing scientific activity. In Edge's (1976) terms, our most general objective is to shed light on the nature of the

'soft underbelly of science': we therefore focus on the work done by a scientist located firmly at his laboratory bench.

(Latour and Woolgar 1986, 27)

Like travelers in the tropics, Latour and Woolgar saw the benefit of recording *there and then* what they saw, even considering it key in their attempt to understand how knowledge is produced. Thus recordings that were made on site, Driver and Martins (2002, 146) write, "were part of a network through which images made on the spot were transformed into authoritative knowledge."

The notion of geographical vision and its influence on geographical science is essential here. As Cosgrove (2008, 5–6) writes:

Vision in the sense of active seeing is inescapable in the practice of geography. (. . .) In the long and now largely superseded meaning of geography as the practice of exploring, reporting and recording the varied surface of the earth—its lands and seas, its climates and environments—eyewitness knowledge and verifying the truth of visual observation were crucial features of geographical science.

However, whether for scientific or artistic purposes, these on-the-spot observers saw the advantage of homogenizing their techniques so that their views might reflect a similar standard or be able to claim truthfulness. While a thorough discussion of the conventions of landscape painting falls beyond the remit of this book, what Driver (2004a, 1) calls the "discipline of the senses," which is required for "the production and circulation of authoritative knowledge" is important here:

This sort of discipline has often been conceived by historians of science in terms of the heightened emphasis on instrumentation within the field of sciences during this period and the impetus to precision which this represents, as for example in the celebrated case of Humboldt himself. While this focus on instruments is in itself necessary and tells us much about the epistemology of contemporary natural science, it is as important to recognize that the making of observations in the field also required the deployment of specific kinds of embodied skill in the production of images and inscriptions—as reflected, for example, in what might be called the "instrumentalisation" of hand and eye.

(Driver 2004a, 1–2)

The discipline that Driver speaks of with regard to recording the tropics is akin to the one set at sea, on surveying ships, and by cartographers. Be it a single longitudinal reference or the agreement on the length of a mile, the need for a standardization of measurement was foremost. This was essential for practical reasons such as cartographic coherence, as well as to give credibility to representations at sea, in the tropics, or even on smaller, national scales. There were various ways in which standardization was achieved, such as fixing units and publishing instruction manuals detailing procedures and protocols to be followed to record observations, and these will be discussed in more detail at a later point. However, this third similarity between the geographical imagination of the tropics and that of the ocean-space is methodologically important.

European encounters with faraway lands in general and the tropics specifically were, initially at least, imaginary ones, constructed through cultural artifacts and representations. From a European point of view, non-European spaces were perceived as ones that stood opposed to Europe. This is the cornerstone of Said's (2003, 1–2) concept of "orientalism" where "the Orient has helped to define Europe (or the West) as its contrasting image, idea, personality, experience." With regard to the tropics, Arnold (1998, 2) writes:

> Calling a part of the globe "the tropics" became a Western way of defining something environmentally and culturally distinct from Europe, while also perceiving a high degree of common identity between the constituent regions of the tropical world. The tropics existed in a mental and spatial juxtaposition to the perceived normality of the northern temperate zone.

In other words, from the seventeenth to nineteenth centuries, "the tropics were being as much created as discovered" (Arnold 1998, 4). Through diaries and journals, the tropics were constructed in an "imaginative geography" or "geography of the mind" that did not necessarily bear much resemblance to any physical reality (Livingstone 1992, 57; Lowenthal and Bowden in Arnold 2000, 6). Indeed:

> The 'tropics' was simultaneously a set of material facts (an area of the globe, characterised by certain climates, peoples and organisms), a set of discourses (ideas and ways of thinking about this part of the globe and its relation to humans), and a set of projects (imperial, bureaucratic, commercial, religious and gendered ways of interacting, in part shaped by local structures, and in part by imperial structures that transcend the tropics).
>
> (Duncan 2000, 34)

In the mind, ideas of tropicality emerge as homogenous and the tropics as a zone that had yet to be specified and physically located. In the same way, geographical imaginations of the ocean-space were also constructed as a unified space, overall standing opposite to land: oceans were perceived as unitary spaces whose principal defining characteristic was that they were not land. Furthermore, cultural depictions of the ocean-space also contributed to the construction of an imaginative geography of the ocean-space. Certainly, in art and literature, until the twentieth century, novels that were focused on the ocean were scarce: the ocean, and, in particular, the Mediterranean Sea, permeated some literatures, such as ancient poetry and the Bible, but it was rarely central. During the European renaissance, the sea regularly appeared as a backdrop in Classical-themed paintings, but it was not until romanticism that it came to full artistic prominence. Many romantic artists and poets were greatly attracted to the sea, idealizing it in the gothic novel genre or paintings. Tightly intertwined with the Alps to epitomize the sublime, the sea became a metaphor for all that is frighteningly beautiful and "the expression of great spirit" (Longinus in Kuiper 1995, 1077). With Victor Hugo and the British romantics, the sea was awe-inspiring and unwelcoming. J.M.W. Turner paints the sea's horrors in impassioned depictions. This sea is romantic, frighteningly stirring, and shrouded in a thick aura of mystery.

With the triangular trade and the Industrial Revolution, Victorian authors, such as Charles Dickens, carried their readers to ports and harbors, describing these places in their mechanical, maritime splendor. Conversely, the sea was simultaneously praised for its restorative properties, as countless protagonists were sent to the seaside in order to heal, as "Europeans discovered the beach, whose water and air they sought for health reasons" (Rozwadowski 2005, 4). Progressively, the sea thus became less evasive, and, therefore, its representations became less blurry. By the time that railroads were able to carry city dwellers to the seaside and as bathing became more popular and socially acceptable, the sea became, if not tamable, then at least less scary or intimidating. It became picturesque, associated with colorful cabins and refreshing air. Yet, essentially, the sea remained secondary: it was never the focus of attention. What people sought was to be by it, not actually involved with it.

However, by this late-Victorian period, a new literature had emerged perhaps best exemplified by the following: Herman Melville's *Moby Dick*, Jules Verne's *Vingt mille lieues sous les mers*, and Joseph Conrad's novels and novellas. In this new genre, whose depictions of the sea were a melting pot of various and diverse images of the sea, the ocean-space becomes a setting for stories and its physicality was placed at the core of novels. These novels represent the sea and are the ocean, to the intents and purposes of their readers. The sea they describe is exciting, "[imbuing] the act of going to sea with new meaning, creating the expectations that generations of passengers and sailors took with them aboard ship" (Rozwadowski 1996, 413). One can be free there, defy gravity and the materialism of cities, and believe in harnessing the elements. With these authors, the ocean is given an image and a voice. Verne's Captain Nemo's description of the sea, for instance, bears witness to this poetic sea. Nemo proclaims:

> The sea is everything! It covers seven tenths of the globe. Its breath is pure and clean. It's the immense desert where man is never alone, for he feels life shiver by his sides. The sea is only the vehicle of a supernatural and prodigious existence; it is only movement and love; it is infinity alive, as one [French] poet said. In effect, professor, nature manifests itself here in its three forms: mineral, vegetal and animal. (. . .) The sea is the vast reservoir of nature. It is by the sea that the globe began, and, who knows, maybe it will finish with it. There is utmost tranquility. The sea does not belong to despots. On its surface, they can still exercise unjust power, fight there, eat each other, make terrestrial horrors. But thirty feet underneath its surface, their dominion stops, their influence is gone, their powers disappear. Ah, sir, live, live within the seas! Here is independence. Here I have no masters. Here I am free.
>
> (Verne 1990, 125)

Nemo's sea is alive, full of potential and excitement. Possibilities, at sea, are limitless, and it becomes possible, in these novels' pages, to navigate under the Suez Canal before it was actually finished or even going underneath Antarctica.[7] At this point, the sea, while still sublime, was now also tantalizing, powerful enough to hypnotize and entrance.

Later, other narratives of the ocean-space would also play a role in shaping a cultural sea. In particular, with the success of large-scale scientific expeditions at sea, such as the HMS *Challenger* expedition and the British Antarctic expeditions, "scientists published accounts of their experiences addressed not only to their scientific colleagues, for whom journal articles and monographs would have sufficiently presented their discoveries, but also to a popular audience with a growing interest in the sea" (Rozwadowski 1996, 429). This was part of a "refashioning of maritime culture," which "reflected growing nostalgia for sea life" (Rozwadowski 1996, 428).

The "imaginative geography" of the ocean-space became constructed through a series of cultural and artistic expressions of the ocean. Certainly, the cultural ocean bears witness to the importance of the maritime "as a shaping fantasy of the cultural imagination as much as in defining the ocean as a deeply historical and radically political space" (Klein 2002b, 10). Like the tropics, the ocean-space was a set of material facts, discourses, and projects, which created the ocean-space's geographical imagination.

However, when thought of in terms of specific locales, it is worth noting that neither the geography of the tropics nor the ocean-space was as uniform as their imaginary geographies perhaps portrayed them. Both would be fine-tuned and began acquiring localized characteristics. Originally, " 'the tropics' meant to Anglo-Americans the tropical parts of their own continent, or that extended region from Virginia to Brazil, which produced such crops as cotton, sugar, and tobacco" (Arnold 1998, 3). However, ideas of tropicality have since been applied, among other places, to most of South America (Naylor 2000a), India (Arnold 1998), Ceylon (Duncan 2000), and the Malay Archipelago (Taylor 2000). Invariably, though, the tropics were in essence " 'a new world' that held out to Europe the prospect of excitement and discovery, a storehouse of untapped wealth and knowledge" and whose particulars could be glanced over (Arnold 1998, 3). With regard to the ocean-space, local characteristics of specific shipping routes, coastal zones, or individual seas might differ from the imagined ocean-space as a whole, but these did not redefine the ocean-space, since it was still geographically imagined in its entirety.

As a uniform space, however, the tropics were imagined in similar ways: opposed to the Western, temperate zones and in turn both paradisiacal and pestilent. The same is true of the ocean-space where, for instance, the seabed, as a space with the ocean-space, also acquired successive contradictory epithets. As Rozwadowski (2005, 70) writes:

> Before the commercial and strategic potential of submarine telegraphy drove investigations, the ocean floor appeared to hydrographers as a violent, rugged

place. Later, with commercial and scientific uses that [mid-nineteenth century] engineers and scientists had in mind for the deep sea, its floor metamorphosed into a flat, quiescent environment safe for submarine telegraph cables.

This particular shift was prompted by a specific event. In the case of the tropics, visions altered for a myriad of reasons.

Cosgrove (2008, 105) writes:

> On the one hand both classical and biblical traditions placed the city at the peak of a hierarchy of environments at the base of which was wilderness, that is the pre-social, pre-discursive space. The marvellous offered a countervailing theme to the monstrous image of savage nature, suggesting that somewhere at the utmost ends of the *oikoumene* were the spaces of perfection and wonder: the dwelling place of heroes beyond the pillars of Hercules, the prelapsarian world of the hyperboreans beyond Thule or the terrestrial paradise of humanity's innocent childhood in the east.[8]

European geographical imaginations of nonurban or uncivilized spaces are thus of two kinds: closest to the city lay the savage and monstrous lands, inhabited by terrible races of hybrid creatures, but beyond these are earthly paradises where gods and heroes dwell. In either case, "beyond the habitable lands lay wilderness, which, whether savage or gentle, was a place that lacked pathways and where human dwelling remained impossible, because its lands were not subject to culture and civilization" (Cosgrove 2008, 105). Certainly, these two visions of the tropics are evident throughout the literature: firsthand accounts and secondary sources alike represent either of these perspectives with equal certitude.

Since Christopher Columbus's initial crossing of the Atlantic Ocean in 1492, the tropics were painted in Europe "as an earthy paradise and a world of natural (rather than man-made) abundance" (Arnold 1998, 3). According to Columbus's accounts, the tropics were

> lands of great natural abundance, alive with luxuriant vegetation and exotic birds and animals, and blessed with perennially warm climates. Spared the cold and hunger of northern winters, humans in the tropics could enjoy easy, year-round subsistence in return for minimal labour.
>
> (Arnold 2000, 6)

This Edenic, full-of-life imagination of the tropics was fueled by explorers' and travelers' accounts that described the landscapes as astonishing, therefore sustaining a "full-fledged myth of tropical exuberance" (Curtin in Arnold 2000, 6). In fact, despite some familiarity with tropical flora, which he had encountered in botanical gardens, upon seeing the tropics, Darwin (in Martins 2000, 24) wrote: "it would be as profitable to explain to a blind man colours, as to a person, who has not been out of Europe, the total dissimilarity of a Tropical view." Furthermore, from the point of view of the naturalist, Darwin was enchanted by the richness of the ecosystem he was visiting. He wrote: "here [the naturalist] suffers a pleasant nuisance in not being able to walk a hundred yards without being fairly tied to the spot by some new and wondrous creature" (Darwin in Martins 2000, 27). Alexander von Humboldt was also influential in shaping the tropics as "a realm of aesthetic appreciation" (Arnold 2000, 8). The journal of his 1799–1804 travels across the Atlantic Ocean and the Americas, *Personal Narrative of a Journey to the Equinoctial Regions of the New Continent*, described a world of "organic richness" and "abundant fertility" (Von Humboldt in Arnold 2000, 8). Accounts such as these fashioned the geographical imaginations of the tropics as wonderful spaces, even prompting some to posit that the Garden of Eden would actually be found there (Livingstone 2000, 93).

Conversely, other imaginations of the tropics perceived them as hellish, an entirely negative space. Specifically, the tropics were the breeding grounds of diseases to which Europeans had little or no resistance: the tropics were "environmentally hazardous and disease ridden," and this very much tainted their image (Arnold 1998, 4). As Arnold (1998, 4–5) explains:

> One factor was the growing evidence of white mortality in the American tropics and West Africa, and the failure of the Dutch, French, and British to establish colonies of settlement there and to reproduce themselves, biologically and culturally, in such alien places.

Beyond the practical problematics that they posed to imperial, commercial, or religious enterprises, these negative discourses of tropicality were closely linked to wider debates on the nature of humanity and were laced with attempts to demonstrate the physiological superiority of the European (Arnold 1998, 5). Indeed:

> the tropics were made to bear a moral message that flattered Europe's sense of superiority while denigrating its alien "other." Although describing the tropics as "nature's garden" might seem to suggest unqualified approval, in an age obsessed with improvement and progress, with racial origins and competitive evolution, there were definite disadvantages to being the denizens of an earthly paradise. This was a primitive world, a land that civilization had shunned. The "primordial" character of tropical landscapes was most forcefully suggested by the vast and seemingly inexhaustible forests. Whether naturalists found the forests of Amazonia grand or gloomy, they understood them to be an obstacle to progress and the advance of civilization.
>
> (Arnold 2000, 10)

Following this, the possibility of finding the Garden of Eden in the tropics faded: as they were emerging as disease ridden and pestilent, the tropics' paradisiacal appeal was disappearing. Later, even, some argued that the physical properties of the tropics would turn into moral decadence and dirtiness, thereby furthering the idea that civilization was not to be found in that distinctly nontemperate zone

(Duncan 2000; Naylor 2000a, 50), leading to an idea dubbed "the morality of climate" (Livingstone in Duncan 2000, 45).

Furthermore, in terms of practicality, the need to deal with "impenetrable jungles with underwood and climbing plants and poisonous or useless fruits" was a real challenge for Europeans (Taylor in Duncan 2000, 38). As well as the physical hardship of the place, Europeans were confronted with a practical reality of the tropics, which invariably differed from expectations. This highlights the difference between what Livingstone calls the "anticipative geography of the tropics" and the reality of everyday life in the harsh tropical environment: there was a "betweenness of [the] place" in that physical reality did not coincide with the geographical imaginary (Duncan 2000, 38; Livingstone 2000, 94). Thus "the eyes of the departing traveller were not quite the same as those of the returning traveller," and their accounts of harsh realities marred Edenic visions (Martins 2000, 21).

However, while these two visions of the tropics are arguably the most prominent ones, there was a third way of looking at the tropics that shaped geographical imaginations: the tropics as "useful" and "exploitable" (Naylor 2000a, 49). This discourse is tightly linked to those of colonization and imperialism, where "Europeans strove to incorporate and subjugate the tropics" (Naylor 2000a, 50).

As discussed earlier, the "imperial eye" by which visions of the tropics were tainted by "an eye of surveillance and control that symbolically takes over or possesses that which it looks upon or over, including places not yet actually taken over or occupied by imperial Europeans" is integral to how knowledge of the tropics was produced (Bell in Martins 2000, 20). European interests in the tropics are necessarily underpinned by such discourses. Yet focusing on the geographical imaginary of the tropics is useful here in terms of understanding how knowledge about faraway spaces is made in the first place, not how these spaces were then used; indeed, I consider the processes that make knowledge about the tropics inasmuch as they are helpful to grasp how knowledge about the ocean-space was made.

What is crucial here is that the discrepancy between visions of the tropics or tropical imaginaries highlights the manner in which the geography of faraway places was produced and how it reached the cultural sphere. In their work, Arnold, Driver, and Martins highlight the convoluted processes through which knowledge about space is produced. Certainly, the superimposition of cultural constructions and conflicting reports make for a multifaceted and amalgamated geographical imagination. Here the tropics serve as an example of a faraway space, which was vastly unknown until recently and upon which was opposed a variety of defining labels. Whether negative or positive, ways of describing the tropics as a space or tropicality as a cultural construction were justified: the tropics were luxuriant at the same time as they were a harsh environment. How these two visions interacted with the cultural imaginations is what is useful to the present discussion. As Driver and Martins (2005b, 5) put it, the tropics were not simply homogenous "screens" upon which European visions could be projected, but rather a complex interaction of exchanges and negotiations. The diaries of von Humboldt and Darwin presented one form of knowledge from a particular viewpoint, namely the naturalist's gaze, whereas private letters, such as the one from seventeen-year-old James Taylor quoted earlier, described a harsher daily reality. These views then appear to have been magnified and came to define the tropics as either completely abundant or thoroughly grueling. From there, tropicality emerged as a two-sided geographical imaginary who's physical and cultural boundaries were blurry. The historical geography of tropicality, as projected on Driver and Martins's screen, was drafted as accounts were being written and mailed home, creating an amalgamated vision, an idea of what this particular faraway place should look like and how it might fit within a wider, cultural cosmography. For the tropics, this meant setting the tropics against the West. With regard to the ocean-space, this signified, in the first instance, making mental space for the world's oceans to exist, in accordance with conflicting reports, visions, and experiences. Indeed, if geography is about knowing the world, the historical geography of faraway spaces, be they tropical or oceanic, is about making sense of these as best possible through networks of communication and knowledge.

Connecting places and networks of trust

Central to the making of geographical knowledge about faraway spaces is the problem of receiving and gathering information about these faraway spaces. Indeed, as Driver (2004b, 73) asks, "How should information about distant places be collected? (. . .) And how far [can] it be trusted?" These two questions are at the core of what Withers (1999, 498) calls the "historical geography of trust" and what Mayhew (2005, 75) examines as the "construction of credibility." This section will discuss particular human networks needed in order to make geographical knowledge at a distance; the following section will examine more specifically the scientific and physical elements of these questions. While these outlooks are artificially divided here, they are certainly in practice wholly intertwined. Indeed, Driver (2004b, 75) writes of the concept of a "culture of exploration" as:

> a useful means of highlighting the ways in which ideas, images and practices of exploration traversed the realms of public culture during the long nineteenth century. It is not that boundaries do not exist between, say, scientific exploration and adventurous travel, the sober and the sensational, or the analytical and the aesthetic. (. . .) The business of the scientific explorer was not always, or easily, distinguished from that of the literary flaneur, the missionary, the trader, the imperial pioneer.

The multifaceted aspect of exploration was evident with regard to the tropics, where tropicality emerged as a compound of scientific inquiry and geographical imaginations; the present separating of these perspectives here is rather to give a more thematically comprehensive overview of the literature.

Shapin (1995, 255–256) asserts, "the credibility and the validity of a proposition *ought to be* one and the same. Truth shines by its own lights" (my emphasis). However, he continues, if "the truth of knowing and the truth of being ought to be the same (. . .) , in practice we can never be quite sure that they are." Locating their work during the Scientific Revolution, Cook and Lux (1998, 179) follow this, writing:

> As many historians have remarked, the new philosophy of the seventeenth century self-consciously privileged the depiction and description of things and events over discussion and debate between antagonists. But new "matters of fact" were often unpredictable, strange, or anomalous, things that could never have been constructed from learned argument. Because they could be discovered only through observation or investigation, the truth or error of reports about new matters of fact could not be judged according to reason alone. Establishing new knowledge about nature therefore depended upon multiple and sometimes ambiguous ways of judging the reliability of matters of fact.

Thus the question of trusting information about faraway places becomes one of trusting the informant. With regard to geography, Withers (1999, 497) writes: "geographical knowledge [in the late-seventeenth century] was based not simply upon mapping, reporting, and direct personal encounter but upon establishing trust and credibility in negotiating social boundaries." Demonstrating the credibility of "matters of fact" was akin to certifying the integrity of the messenger of any particular matter of fact (Shapin and Schaffer in Cook and Lux 1998, 179). This was often as difficult as acquiring the information in the first instance.

Cook and Lux (1998, 179) discuss one particular method that would ensure the reliability of information: networks and "strong social bonds among communities of investigators." For instance, by organizing communal viewings of certain phenomena and experiments, these gained veracity through a simple law of numbers. If several socially established persons claimed to have witnessed the same event, the event was deemed likely to be true and usable as a foundation of knowledge. If such a viewing could not be organized, a strict dictation of methodology could be distributed among a tight circle, and the use of these approved methods would vouchsafe credibility. In the British fleets, this is exemplified by the publication of a series of *Manuals of Scientific Enquiry* that were published by the United Kingdom Hydrographic Office and distributed to ships (see Herschel 1949). These clearly and precisely outlined procedures for operations such as dredging or sounding in an attempt to homogenize procedures.

However, overall, the preferred guarantor of credibility was social status. Cook and Lux (1998, 179) write:

> whether the gentlemanly status of the main body of fellows of the Royal Society, the aristocratic status of an investigator's patrons, the social legitimacy of a religious order or university faculty, or the etiquette and honour of courtly life, social prestige is said to have established the credibility of matter of fact asserts by members of a group.

Indeed, the higher classes had monarchic "pretensions to absolutism and the desire to control ideas and events," despite seldom having firsthand experience of the knowledge they were monopolizing (Cook and Lux 1998, 180).

The inherent contradiction of using social status as a barometer of credibility is that these higher classes did not travel themselves: they relied on "invisible technicians" who were not members of the elite (Shapin in McCook 1996, 179) to act on their behalf. At the same time, though, as Cook and Lux (1998, 180) note, these nontraveling gentlemen scientists would have been inclined to think that "simple people were not prejudiced by theory or clever enough to lie credibly." Therefore, these simple people's credibility was guaranteed because they were simple.[9] Furthermore, it was the journeys themselves that bestowed upon these simple people the credibility that they needed: "travel helped to shape three things: learning, experience, and judgement" therefore serving as a warranty for the nature of the knowledge brought back (Cook and Lux 1998, 185). Finally here, the actual traveler, however apparently simple, was in the unique position of being able to establish direct relationships with those encountered. These new contacts, presumably from a range of classes of society, became part of the network of credibility as "acquaintances [began] vouching for the credibility of others," thus expanding, if perhaps also weakening, this network (Cook and Lux 1998, 190). Certainly, personal relationships were instrumental in the furthering of knowledge acquisition as a wider, vetted network allowed for more and more knowledge to be exchanged.

Among the acquaintances that the traveler could make was with locals of the places that were being explored. Considering the making of geographical knowledge about the Scottish Highlands, Withers (1999) examines the networks of local knowledge and their place in geographical knowledge. Here two factors are at play: the credibility of the sources and the legitimacy of their knowledge. In this case, it transpires that establishing credibility was at the discretion of the patron. For example, one Robert Wodrow was interested in the Western Highlands and, through a complex network of letters and questionnaires, sought to increase geographical knowledge of the area. His network had substantial local knowledge that Wodrow wanted to integrate into his project of Scottish geography. Wodrow's correspondents in the Highlands were mainly known to him personally and often Church of Scotland ministers. Withers (1999, 505) writes:

> [Wodrow's] reliance on such men is understandable: there was almost no one else in the region who might be judged reliable or capable of finding things out and reporting back. Gaelic speakers themselves, they no doubt drew on native vernacular knowledge as legitimate knowledge. (. . .) Wodrow was prepared not only to accept local knowledge as legitimate but to trust his informants, laymen and ministers alike.

What this passage highlights is the importance of negotiating both the networks of knowledge themselves and the nature of the knowledge being transmitted.

The dichotomy between local knowledge and administered, or official, knowledge is noteworthy with regard to knowledge about the ocean, and especially in the instance of the European discovery of the Gulf Stream, which I discuss in Chapter 4; what is crucial here is the importance of the credibility of informants through networks of trust. Indeed, Wodrow chose to trust otherwise unconventional sources because they were part of his personal network of acquaintances.

Following the direct, in-person establishment of credibility, the making of knowledge also needed to be transmitted and transported within these networks. The principal method by which this was done was letters, and Cook and Lux (1998, 202) comment that, in this respect:

> the "scientific community" of early modern Europe operated in a manner little different from other communities of people, whether scholars, diplomats, merchants, or any others who travelled long distances and followed up their meetings with correspondence.

Letters provided an informal way of communicating within a predefined and established network and transmitted knowledge among its ranks in an easy and reliable manner. In particular, in eighteenth-century Europe, the formalized network of the Republic of Letters provided a useful platform for discussing matters of scientific interest. As Mayhew (2005, 74) writes:

> At the level of professed ideals, the concept of the Republic of Letters was fairly simple. Scholars would create an egalitarian world among themselves in which scientific opinions could be exchanged without the rancour of national, historical or other barriers clouding their judgement.

The significance of such a platform is manifold. First, the Republic of Letters operated beyond national cultural or historical settings. Mayhew (2005, 74) comments on the "impressive degree to which eighteenth-century scientists ignored war and xenophobia to exchange ideas." The network of the Republic of Letters enabled scientists to strive for "cosmopolitan and universalist ideals that were [otherwise] undermined by national antagonisms and religious schisms"; this provided science with a network of credibility that operated beyond usual social realms (Mayhew 2005, 74). Connected by "ink, both on the printed page and in the written letter," the Republic of Letters was an organized network of informants that was, following Anderson, "an imagined entity, held together by ties assumed and asserted through non-personal modes of interaction" (Mayhew 2005, 76).[10]

Furthermore, belonging to the Republic of Letters immediately increased one's own personal network of acquaintances without jeopardizing the nature of these acquaintances' credibility. Perhaps the links were weaker overall than they would have been through personal, direct networks, but the increased pool of informants and the overcoming of spatial distances likely compensated for this.[11] The driving force for the enterprise was a utopian one that sought to overcome distances while trying to least compromise the quality of the exchanges. As Mayhew (2005, 92) concludes, the "ideal of open communication was a powerful one which did lead scientists and scholars to open their eyes to a far-flung community and to see all as sharing an essential kinship." In this context, the Republic of Letters serves as an example of trying to overcome the problematics of establishing credibility while making geographical knowledge over a large area of the globe.

The organization of the Republic of Letters mirrors that of certain professional networks, and especially guilds and learned societies, but of particular interest here are networks that existed among sailors. Indeed, military and merchant navies formalized their information gathering through logbooks and cartographic practices, producing homogenized catalogues of geographical information about faraway spaces. In the case of the East India Company, for example, Ogborn (2002, 162) writes:

> [the journals that the company required the captains, masters, and pursers to keep] were to be compared to one another during the voyage "soe as a perfect discourse may be sett downe." As the written record of the voyage, these journals were to be both a guarantee to the adventurers in London of the performance of what was set out in the commission and, where possible, the foundation of succeeding voyages by providing knowledge of winds and shoals, useful ports and places of refreshment, good routes to take, supposedly friendly or treacherous peoples, good commodities and markets, and the extent of Portuguese and Dutch power as well as the orientations of Asian polities. These journals were collated, archived and used by the company in increasingly systematic ways in order to provide accurate "navigational" knowledge for subsequent voyages.

By virtue of being associated with a professional body, the logs or maps were imbued with instant credibility, exemplifying a professional network of trust. This was, in part, thanks to the work of Alexander Dalrymple, then hydrographer to the British Admiralty, who worked toward consolidating and standardizing "the compilation and production of charts used by the British merchant and naval vessels" (Akerman 2006b, 12).[12] Therefore, when Matthew Fontaine Maury sought to make a comprehensive map of large-scale water movements of the world's oceans and turned to ships' logs, these were, in their nature, deemed trustworthy of providing reliable information. (This will be discussed further in Chapter 5.) The network in which ships' logs were made and used, in effect, guaranteed their credibility.

Lastly, the Republic of Letters highlights the use of writing as a technology of knowledge and, in particular, the issues of how writing travels. Ogborn (2002, 156) writes:

> within the history of science the production of knowledge is increasingly understood by paying close attention to the constitutive role of objects, instruments and modes of transport in the making of knowledge around the world.

Therefore, the Republic of Letters is essential as the materiality of the letters themselves can be followed within a formalized network. Using the "materiality of the texts," Ogborn (2002, 156) examines the physical trail of letters in the seventeenth century in the East India Company. Following Greenblatt (in Ogborn 2002, 157), the letter becomes a "mobile technology of power." Indeed, as Naylor (2000b, 1625) points out, "it is often through (. . .) mundane objects that networks of capital, communication, and control are built in": therefore, the significance of the Republic of Letters is derived from a wide network of credible sources and the immateriality of this network, except for the letters themselves. As Harris (1998, 273) writes, the "geography of knowledge also means the dynamics of travel": the manner in which exchanges travel in their networks emphasizes the materiality of the exchanges themselves. In this particular long-distance network, the dynamics of the material letters become as important as the letters themselves or, indeed, "travel writing was of far less significance than how writing travelled" (Ogborn 2002, 168).

The final aspect of credibility that is useful here is what McCook (1996, 177) calls the "legitimisation of evidence." This links to the following section in this chapter that discusses "venues of science," but the particular example examined here is concerned with the broader Victorian social scene (Livingstone 2003, 17). McCook examines the case of Paul du Chaillu, a French-American explorer who brought back gorilla skulls to England in the mid-nineteenth century. These skulls became the focus of attention in a series of heated debates about the biological nature of the gorilla and its closeness to the human (McCook 1996). Du Chaillu's credibility, however, became questioned when his travel narratives appeared to have discrepancies. This led to debates about "the process by which objects collected in the field became scientific evidence" (McCook 1996, 179). The problem with du Chaillu was that he based his credibility on his experience (he had spent a considerable amount of time in Gabon), whereas his peers judged authority on status. As noted earlier, "in the nineteenth century the only people who had the authority to produce scientific knowledge were the self-selecting 'gentlemen of science'" (McCook 1996, 182). As these gentlemen rarely traveled, they had to rely on "invisible technicians" whose observations and samples had to go through the process of "intellectual legitimation" (Shapin in McCook 1996, 179 and 183). Indeed, "confirmation by a scientist was essential for an object of natural history to become scientific knowledge," and du Chaillu's ambiguous status posed problems in this otherwise well-defined process (McCook 1996, 184). During the episode that became known as the "Gorilla Wars," one reviewer labeled du Chaillu as untrustworthy because "he took no observations, either astronomical, barometrical, meteorological, or thermometrical," while recognizing him to be "an energetic and active explorer" (in McCook 1996, 189). This highlights "the subtle hierarchy between the different categories of traveler": at stake was du Chaillu's "reliability as a witness" (McCook 1996, 189 and 191). Not only did the "facts (. . .) not 'speak for themselves' and du Chaillu (. . .) not speak for the facts," but du Chaillu did not even benefit from "Englishness and rectitude": his credibility, therefore, was never fully acquired in social circles (McCook 1996,

193–194). Whatever combination of criteria was used to establish trust between gentleman scientists and those who went to the field were rarely connected to the nature of the science performed in the field. And indeed, while "out of necessity, metropolitan scientists accepted reports and collection from people who were on the social margins of the scientific world," these were seldom judged on scientific principles (McCook 1996, 196). Until the notion of "trained observer" was introduced and standardization became widely adopted, credibility based on social connections pervaded all aspects of science (McCook 1996, 197).

Returning to Driver's initial questions, there were several ways to overcome the problem of making geographical knowledge, especially a number of social formations. These notions are essential to understand how knowledge could be made at a distance; however, at sea and aboard ships, the particular social mores that governed the establishment of credibility was further complicated by specific rules that governed life aboard ships. These inevitably shaped the nature of knowledge that was being made about space.

Historical geographies of science

STANDARDIZING AND MEASURING SPACE

Livingstone (2003, 1) writes:

> Scientific knowledge is made in a lot of different spaces. Does it matter where? Can the location of scientific endeavor make any difference to the conduct of science? And even more important, can it affect the content of science?

Livingstone's position is that the answer to these questions is "yes," noting, however, that "the idea of a geography of science runs counter to our intuition. Science, we have long been told, is an enterprise untouched by local conditions. It is a universal undertaking, not a provincial practice" (Livingstone 2003, 1). Yet locality is at the core of the making of science and provinciality brings to the fore what Gieryn (2002, 113) calls "the paradox of place and truth" or what Connery (2006, 495) examines as "the locational character of knowledge." Furthermore, the space of the laboratory raises conceptual and practical issues, especially within the setting of ocean science. Indeed, these two questions relating to sites of knowledge production challenge science's universality, signifying that science has a geography and takes place in a multiplicity of spaces. Among others, Thrift, Driver, and Livingstone (see Driver et al. 1995) have written about

> the bewilderingly diverse array of sites of scientific production (laboratories, lecture halls, observatories, museums and map rooms, for instance), a range of sites where scientific knowledge might be constructed at a distance (scientific societies and associations, corresponding societies, lecture tours, exhibitions, scientific publications), a cornucopia of technologies of knowledge

disseminations and presentation, and a set of spaces where these knowledges can be legitimately collected.

(Naylor 2002, 495)

Naylor (2002, 495) expands this list by citing more imaginative sites of science that geographers, historians, anthropologists, and sociologists have written about: "physic gardens, field sites, museums, scientific societies and institutions, the home, the body and even the pub."

Yet, crucially, wherever science takes place should not affect what it produces. For knowledge to be credible and scientific, it has to be universal, "not bear the marks of the provincial": "science that is local has something wrong with it" (Livingstone 2003, 1). But, as any activity, science has to take place *somewhere*: "where else could science take place but in places, and how else could it travel but across spaces?" (Shapin in Naylor 2005a, 2). In other words, as Gieryn (2002, 113) writes:

All scientific knowledge-claims have a provenance: they originate at some place, and come from there. However, as they become truth, these claims shed the contingent circumstances of their making, and so become transcendent (presumably true everywhere, supposedly from nowhere in particular). Turning the argument around: scientific claims are diminished in their credibility as they are situated somewhere, as if their truthfulness depended upon conditions located only there.

The contradictory fact that universal science has local origins is at the core of the problematics of the geographies of science.

Furthermore, localizing science, or thinking about science locally, goes beyond placing laboratories or observatories on a map: it acknowledges that

science should be treated like any other form of knowledge; that is, as "a cultural formation embedded in wider networks of social relations and political power, and shaped by the local environments in which its practitioners carry out their tasks."

(Naylor 2005a, 1)

Science is, in fact, a practice "utterly grounded in its social and spatial—not to mention temporal, political and economic—contexts" and "the power of science is not due so much to an unmediated access to the truth as to an unprecedented control over space" (Naylor 2006, 408). Following Latour, Barry (1993, 463) writes:

science is primarily a form of knowing and acting at a distance—of constituting a relationship between esoteric and localized laboratory practices and a network of distant objects. The power of a scientific argument or a measurement is not determined by its truth, but rather judged in terms of its capacity to act across space and time—to mobilize a network of social and technical actors.[13]

This specific question of using science to control space and the idea of science extending over long distances to manage knowledge is useful here with regard to the making of knowledge away from centers of calculation. Indeed, networks of credibility as discussed earlier sought to make the local obsolete, as the practice of science over space was guided by the standardization of measurements. Porter (in Naylor 2006, 408) talks of the "significance of quantification as key to the movement of knowledge as 'it promotes the fixing of conventions, the creation of stable entities that can be deployed across great distances.'" Standardization then emerges as the panacea for implementing coherence across these distances. It is thus:

> the prerequisite for conquering space—the space between the field and the center of calculation, and the space between nature and language. Only by using agreed-on, standards could knowledge be relieved of the burden of parochial judgement or fickle memory.
>
> (Livingstone 2003, 176)

Only with standardized measures and units could knowledge begin to mean the same thing in the field and back in the laboratory.

With regard to cartography and oceanic exploration, for example, the use of a common standard for an original line of longitude would ensure the coherence and transferability of coordinates and a general agreement on where places were in relation to one another. Also, consensus on the length of the mile was essential in measuring how far places were, or even the size of the earth. Nevertheless, national interests were more powerful than standardized common sense, and, therefore, maps operated on different lines of longitude until the 1884 conference in Washington, DC, held for the purpose of agreeing on a prime meridian.[14] With regard to measures, however, the disagreement between imperial and metric standards is still very alive, as it has been since the meter was introduced in 1675. In fact, the duel between the meter and the imperial yard epitomizes the discussion over natural or artificial units. Livingstone (2003, 176–177) writes:

> debate raged over whether standard units were natural or conventional, divinely sanctioned or humanly generated. In rigorously championing the cause of the imperial yard over against the French meter, advocates variously argued that the yard had the backing of tradition, that it was appropriately related to the obvious standard distance of the earth's polar axis, that it gave long-standing expression to Britain's commercial superiority over the French, and that it enjoyed divine warrant by virtue of its direct connection with the dimensions of the Great Pyramid of Giza. The British yard simply must triumph over the atheistic mensural system of French republicans! In France, the defense of a "natural" system of measures was bound up with a revolutionary ardor for obliterating all traces of monarchial caprice. By

determining that the meter should be one ten-millionth of a quarter of the great polar circle (that is, of the distance between the North Pole and the equator), the needs of both science and ideology could be met.

Establishing measurement itself comes with difficulties, as units themselves are laden with the symbolism of place and political agendas. Standards and units, and their respective histories, also emerge as, to return to Livingstone (in Naylor 2005a, 1), "a cultural formation embedded in wider networks of social relations and political power": standards are in fact laden with particular messages.

Similarly, the process of measurement also raises questions. As seen earlier, there were social networks in place to ensure the credibility of information. But standards of measurement were another way of ensuring the unbroken transmission of information or data from the field to the center of calculation. Livingstone (2003, 175) discusses the use of the Munsell color system to identify colors reliably at a distance. Similarly, the development of the Beaufort scale in 1805 was crucial in homogenizing observations of wind speed. By recording phenomena according to an established, standardized scale of measurement, knowledge about a particular phenomenon become transferable and transportable. Likewise with experiments: if a specific set of rules is followed, the experiment is duplicable, and the actual place in which the experiment is run becomes irrelevant. Measured science is thus dislocated: how things are measured, and using what standards, is significant in both the making and the transport of knowledge. The history of science begins to emerge as being closely tied to the history of measurement and the standards of units.

However, as Barry (1993, 467–468) writes:

> Empirically, there is a need to recognize that the history of measurement must be understood as equally a history of the failures of measurement; a history of the phenomena which remain unmeasurable; a history of the incompetences, and passive resistances of scientists and lay people to the exacting requirements of measurement techniques; a history also of the degree to which the attempts to measure have failed to meet the economic, political and moral demands to which they have been tied.

Like standardization, the precept of measurement suggests neutrality but is, in practice, rooted in cultural and social discourses. Despite these shortcomings, the late-eighteenth and nineteenth centuries (a period sometimes dubbed the Second Scientific Revolution) was characterized by "an avalanche of printed numbers" that stood witness to the tight entwinement of measurement, science, and knowledge (Naylor 2006, 409). Furthermore, the urge to standardize science expressed itself in Britain through the passing of a number of official acts. Naylor (2006, 412) notes that

> in Britain alone a number of acts and inspectorates were established through the 1820s, 1830s and 1840s that required the quantification and standardization

of data, for instance, the Weights and Measures Act of 1824, the Statistical Department of the Board of Trade (1832), the Factory Inspectorate (1833), the Registrar-General (1837), the Observatory of the British Association of the Advancement of Science at Kew (1842) and the Excise Laboratory (1842).

These acts were to ensure that across Britain and beyond, measurements would be homogenous, with no room for locality.[15]

One specific example of making science in which questions of locality, measurement, and standardization are especially central was nineteenth-century meteorology, a discipline whose emergence shows similarities with that of oceanography, with which I will draw parallels. As a branch of science that is concerned with a phenomenon occurring in a specific place in relation to similar phenomena everywhere, meteorology highlights the tensions between the delocalized scientific ideal and the intrinsically local processes being studied. Meteorology exemplifies this by studying weather phenomena from two points of view: specifically for the benefit of localized forecasting and more generally, where the local is only part of a larger, atmospherically multilayered, and completely interdependent system. Nonetheless, as a scientific enterprise, meteorology, "perhaps the ultimate example of the attempt to create universal values, was not removed from local circumstances" (Naylor 2006, 409). Schaffer (in Naylor 2006, 409) notes, "the issue of place was crucial," both for the making of standards and as part of the nature of the work itself. Indeed, the choice of where local or national standards would be measured is essential to the collection of meteorological data: Should an observation site show averages or extremes or something else? In this, meteorology is akin to oceanography: the study of the ocean-space necessitated new scientific standards, but the question of where and how these standards would be established was problematic. Indeed, in such a vastly unknown space, how would it be assumed one chosen standard would be useful or applicable for the whole space?

Like oceanography, meteorology is primarily based on widespread collection of specific types of data over large spaces. These data, while geographically located and limited in quantity, are useful principally in relation to other similar datum, and these, when put together, provide a general view of phenomena over a vast area. Nonetheless, both disciplines stem from localized practices: the "meteoric tradition" was "a highly localized and idiosyncratic activity," while oceanic journeys were aiming for a destination. Naylor (2006, 413) describes the methods of three Cornish weather observers working in the nineteenth century:

[Lovell] Squire's early registers contained daily weather information, including maximum and minimum temperatures, quantity of rain, direction of wind and the height of the barometer in the morning and evening. Meanwhile, Jonathan Couch's registers were in the form of monthly accounts and contained maximum and minimum temperatures, and then only more qualitative information on weather type. In turn, Mr Corbett's register of the weather at Pencarrow, Wadebridge, contained monthly mean maximum and minimum temperature, monthly barometric averages, "Average Degree of Dryness"—taken at 1 p.m.—and "Average Quantity of water held in solution by the atmosphere."

However, unlike at sea and on ships where most actions were, from the outset and for various cultural or technical reasons, regulated, behavior in meteorological observatories was not similarly ordered (Rozwadowski 1996, 410).

For meteorology, therefore, reliance on the aforementioned acts of law to guide the observatories' work was essential. As Janković (in Naylor 2006, 412) argues

> the early nineteenth century marked a shift away from the provincial meteoric tradition with its descriptive and idiosyncratic reports of extraordinary atmospheric events, and towards a collecting endeavour based on standardization, quantification and synchronization. The qualifications required of the meteorologist also shifted from the place-based experience and authority of the provincial cleric-naturalist to the expertise-based metropolitan specialist, who gave little regard to local information.

This particular change exemplifies the delocalized nature of science here, impressing that, somehow, the local is useful only if it is networked back to a national science grid. It is through local stations, dispersed across the country, that national institutions "extended their influence over a national space": science was made national science through local science (Naylor 2006, 431). Therefore, Naylor (2006, 433) contends

> we should treat nineteenth century British meteorology not as an inevitable march towards a standardized national weather but as a set of practices that extended unevenly across a physical landscape, that actively constructed geographies of centre and periphery, and that relied on a set of social and intellectual relations.

The same is true of oceanography: the ocean-space became known through a series of local sites and observing stations that all together, and by reporting back to a center, combined to give a full picture of the ocean-space.

From this point of view, standardization emerges as a "technology of circulation" in the making of knowledge, an unquestionable point of reference that negated interpretation and memory lapses (Powell 2007, 322). These matters were unequivocal advantages of this methodology, comparable only to those of photography and mechanical reproduction during the same period. For these reasons, the Royal Meteorological Society welcomed the use of photography as an "expert witness" for weather observations (Livingstone 2003, 167). Livingstone (2003, 167–168) writes:

> Because the development of meteorology required collecting data from a widespread network of spectators, practitioners embraced photography as a

reliable, ever-present, and untiring observer that could catch and preserve things the naked eye could not even detect. Photography promised to ensure that the very trustworthiness that was so hard to establish from the reports of lay eyewitnesses. It could act as a sieve to separate fact from fiction, information from imagination.

Photography was a further means of standardizing ways of looking at phenomena, though it was not entirely unproblematic. Because its applications to oceanography are limited on a surface level and unavailable underwater until the twentieth century, I will not dwell on the medium here, only noting that the enthusiasm that photography generated highlights the importance of a neutral observer that produced information that could be transmitted over long distances as a panacea for homogenous knowledge making.

Certainly, standardization of measures did not erase locality and provinciality in the sciences being enacted in the peripheries. The introduction of common measures across a nation often had other agendas than just science for science's sake. For instance, Ogborn (1998, 289) writes about the close connection between "the barrel and the warfare state" (referring here to the container, not a synecdoche for a gun). He notes the tight relationship between the need to have a national barrel so that taxes could be levied accordingly, highlighting the interactions between and within the national network. In this case, "it was not the centre that was important but its relation to the network" (Ogborn 1998, 306). Measurement in various places thus became an important geographical tool, in spite of the complexity of making scientific geographical knowledge. Withers (2006a, 714) comments on "the basis for geography's contemporary self-image as a science." Mayhew (in Withers 2006a, 714) writes that it "was, then, the character of geography as knowledge rather than the way in which that knowledge was verified, which led to the epithet 'scientific' being applied to the subject." Considering the mental implications of standardization and measurement within the development of geography as a discipline is useful because it highlights a different perspective from which to conceptualize geography, though it does not fall within the scope of this book to explore this further. Here specific examples of the standardization of knowledge production about the faraway spaces in general and the ocean-space more specifically are one way of attempting to make known the vast oceanic expanse. The main argument here is that standardization is important not simply because of its direct impact on knowledge production but also significantly within a wider cultural setting, as the choice of units, a prime meridian, or the size of a barrel is part of a wider enterprise. The fixing of space on small scales and maps is redolent of how space might be controlled at a distance, and this is intrinsic to knowledge production.

Science on the ship

Laboratories, as physical spaces, are governed by practices that shape the knowledge that emerges from them and are themselves conjectured from an assemblage of exterior factors, both physical and cultural. Latour's scrutiny of laboratories has "shown how scientific practices locally articulated depended crucially on the utilization of information gained from afar and on the ways in which hitherto unfamiliar events and peoples were represented, portrayed, and classified" (Livingstone and Withers 1999b, 17). They are "centers of calculation [that construct] manageable, local representational spaces," or, following Naylor, metonyms for the field site (Latour in Livingstone and Withers 1999b, 17; Naylor 2002, 501). Inside the laboratory, Carter (1999b, 311) writes, "our knowledge of the world (. . .) stands in relation to the unknowable world as a parallel universe might stand to our own: perfectly fitting at every point but created according to entirely different principles." Livingstone (2003, 12) writes:

> Consider the laboratory as a critical site in the generation of experimental knowledge. Who manages this space? What are its boundaries? Who is allowed access? How do the findings of a laboratory's specialist space find their way out into the public arena? Attending to the microgeography of the lab—and a host of other similar spaces such as the zoo, the botanical garden, or the museum—takes us a long way in toward appreciating that matters of space are fundamentally involved at every stage in the acquisition of scientific knowledge. What is known, how knowledge is obtained, and the ways warrant is secured are all intimately bound up with the venues of science.

The laboratory, though completely defined by its space and what it contains, must be, in essence, placeless: by nature, it cannot bear influences of outside space lest this alters the science that takes place within. As seen earlier, science that is located is nonscientific: an experiment conducted in a laboratory in Paris, France, is not to bear traces of its French location and has to be replicable in a similarly placeless space anywhere else on earth. As such, this "'other' space is the one in which the objects of science appear. It is the space in which such entities as 'laws,' 'cells,' 'genes,' 'particles,' 'atmospheric pressure,' and 'mental illness' are made manifest and represented" (Ophir and Shapin 1991, 14). This laboratory space is marked by its innate separateness from the rest of the world, following a

> long-standing tradition in the West [that] retiring from society was a precondition for securing knowledge that was of universal value. (. . .) Ironically, to acquire knowledge that was true *everywhere*, [one] had to go *somewhere* to find wisdom that bore marks from *nowhere*.
>
> (Livingstone 2003, 21)

The laboratory thus offers, as Shapin (in Powell 2007, 312) writes, a "view from nowhere." The space of the scientific vessel adds even a further dimension to this inherent contradiction: indeed, ships conduct studies in moveable laboratories whose positions at any given time are unique. Furthermore, samples that are brought inside the laboratory space are themselves extracted from their original spaces, be they terrestrial or benthic, signifying that their study in the regulated space is at least twice removed from reality, as such. The "'disembedded'—extracted from its place of origin" nature of the ship's laboratory space is then truly extreme, as it displaces both what is being studied and the space where specimens are studied (Livingstone 2003, 29).

Aboard ships, the detached nature of laboratory practices were accentuated due to the physical constraints of the ships themselves. Certainly, the ship itself is an ambiguous space. MacDonald (2006, 642) comments on its " 'placeless' mobility." Casarino (in Ryan 2006, 589) calls it the "most ancient and most modern of spaces." Rozwadowski (1996, 409 and 412) talks about the ship as a "small world" that is "physically and socially alien." She writes:

> From at least the seventeenth century, ships' crews were international assemblages, made up of men caught, as Marcus Rediker put it "between the devil and the deep blue sea." Well into the nineteenth century, the sea was widely held to be an appropriate calling for the poverty-stricken, orphaned, criminal and insane.
>
> (Rozwadowski 1996, 412)

Furthermore, when scientists embarked on ships, there were often tensions between the ship's crew and the scientists. This was the case both aboard Captain Cook's HMS *Resolution* and Captain Nares's HMS *Challenger*, which will be discussed in a moment. For Ryan (2006, 591), it is a site of "social geography," which was "maintained by a steady rhythm of daily rituals and routines, such as the regular church services (always two on Sundays) in the saloon, the only adequate excuse for absence being seasickness." Ogborn (2002, 161) seeks to understand "the ship in three ways: as a material space, as an accounting space and as a political space." For him, the ship's materiality is essential as a technology of power that transports letters, goods, and orders across the world's oceans. Certainly, ships are, as Ryan (2006, 580) notes, "not merely vessels for navigating the surface of the sea. They also shape the ways that cultures imagine and represent the sea. Indeed, any account of historical geographies of the sea necessarily involves thinking also about ships and the spaces on board ships." As seen earlier, the seaborne perspective of travelers arriving in the tropics was instrumental in shaping ideas of tropicality. Here, however, the space of the ship that will be discussed is that of the ship as, following Sorrenson (1996, 221), a "scientific instrument."

Sorrenson (1996, 221) writes that:

> questions about the globe—its shape, geography, and topography, its magnetic, meteorological, and atmospheric properties, its oceans, winds, and tides—could only be answered through actual voyages that physically left behind the laboratories, libraries, studies, and scientific societies of metropolitan centers of empire. Likewise, discoveries about the contents of that globe—its peoples, animals, plants, soils, rocks, waters, and airs—required that a range of scientific instruments be operated by skilled observers in the field, whether that field lay nearby or at the extremities of the known world. One such instrument was the ship.

Thus the ship was more than just an interchangeable carrier of goods: it was specific, it "left traces in the maps" and gave "superior, self-contained, and protected [views] of the landscapes and civilizations" (Sorrenson 1996, 222).[16] In fact, Sorrenson (1996, 222) continues:

> A ship that conducted a voyage of scientific discovery was never *merely* a vehicle that transported investigators to observe mundane new worlds, anymore than a telescope was *merely* a vehicle that transported images of heavenly new worlds to an observer. Just as the telescope expanded the science of astronomy and allowed astronomers to explore new worlds and make images of them, so too did the ship for geography and geographers.

Therefore, the ship was, first and foremost, the opportunities that it represented: as a technology, it made it possible to connect places that were otherwise separated by the impenetrable ocean-space. In contrast with armchair geographers or gentlemen scientists, exploring geographers could obtain facts from the field instead of simply theorizing it. In particular, such discussions as those pertaining to the existence or not of a southern continent could only be proved with a ship, on an expedition. Indeed, "with their ships and globally extensive voyages, these sailing geographers could themselves prove or disprove the existence of phenomena that armchair philosophers and laboratory-bound researchers could only speculate about" (Sorrenson 1996, 223). The ship, as such, enables geography to be, as Louis de Bougainville (in Sorrenson 1996, 223) put it, "a science of facts," rather than one of speculation.

Yet the actual ships, such as the one that "would carry Cook around the world to make islands and unmake continents," were, in essence, no different physically from those that carried goods along the coasts of Britain (Sorrenson 1996, 226). While they might have been adapted under deck to accommodate civilian crew or chosen specifically for particular features, their peculiarities lay in what was happening in these spaces. They were primarily "instruments of geographical discovery," but the scientific nature of geography turned them into "floating laboratories" (Sorrenson 1996, 227; Beaglehole in Sorrenson 1996, 227).

Sorrenson (1996, 227) writes:

> Scientific ships, more generally, also placed naval officers, astronomers, naturalists, and artists on the land where they wished to make observations, or allowed them while at sea to make a wide range of maritime observations using sophisticated mathematical, philosophical, and optical instruments. However, these ships were more than just vehicles or platforms for observers and instruments; they shaped the kinds of information observers collected.

Thus the space of the ship (or space-ship) actively shapes the nature of the observations and, therefore, also knowledge about the geographical spaces being observed. This follows Galison (in Powell 2007, 315) for whom "the buildings of science literally and figuratively configure the identity of the scientist and scientific fields." As Sorrenson (1996, 236) concludes, "As scientific instruments in their own right, they mediated the complex interplay between representation and reality that lies at the heart of eighteenth-century geography."

Furthermore, as well as providing "a superior, self-contained, and protected view of the landscapes and civilizations," the ship is also a space to study these landscapes and civilizations (Sorrenson 1996, 222). In particular, some ships were equipped with laboratory space that provided room to analyze samples of earth or animal specimens collected on land or at sea. This onboard laboratory was a space of science within an instrument of science, and the superimposition of two spaces of science, a vehicle of science and a location of science, created tensions aboard ships, in particular between the vessels' crews and scientific personnel. In fact, the sharing of space aboard ship was at the core of the experience of making knowledge at sea. As Rozwadowski (2003, 1) notes:

> Oceanography as it emerged in the last quarter of the nineteenth century focused the attention of many scientific fields—physics, chemistry, biology, and geology—on the project of understanding the oceans. Its practitioners did not share a common set of intellectual questions, nor do they today. Instead, they shared the experiences of boarding vessels, meeting sailors, and wrestling with the maritime gear used to retrieve data or specimens from the restless sea.

Part of this shared experience was negotiating the maritime establishment, which was laden with tradition, folklore, and technology. Thus Charles Wright (in Rozwadowski 2003, 2), the botanist on the North Pacific Exploring Expedition (1853–1855) complained that "the majority of the [officers'] mess have a most sovereign contempt for science and no esteem for its devotees." One reason for the tensions between crew and scientists was that the latter sometimes imposed more work on the former: dredging in particular was very labor-intensive, and while excitement initially accompanied the event, when it became routine, the novelty wore off. Henry Moseley (in Kunzig 2000, 34), a naturalist on the HMS *Challenger* expedition, writes:

> At first, when the dredge came up, every man and boy in the ship who could possibly slip away, crowded round it to see what had been fished up. Gradually, as the novelty of the thing wore off, the crowd became smaller and smaller at the critical moment, especially when this occurred in the middle of dinner-time, as it had an unfortunate propensity of doing. It is possible even for a naturalist to get weary even of deep-sea dredging.

For crews, dredging generated a disproportionate amount of work for few results: a successful day's result yielded nothing but inscrutable ooze. For sailors, dredging rapidly became "drudging" (Rozwadowski 1996, 417).

Another tension was derived from the superimposition, in the closed space of the ship, of civilian and military persons. Traditionally, a strict hierarchy redolent of the military governed the space within the ship: "space, rank, and behavior were closely related" (Rozwadowski 1996, 415). In Chapter 4, I will discuss this briefly with regard to Edmund Halley aboard HMS *Paramore*. Certainly, as ships had to be reorganized in order to make room for scientists and their equipment, the fine balance of this hierarchy was broken: "scientific work (. . .) took over space originally dedicated to the vessel's military function" (Rozwadowski 1996, 415). In fact, for the first time aboard HMS *Challenger*,

> space was allocated on the principle that the scientific work was equal in significance to the navigation and surveying work. The unprecedented step of dislodging the captain from his traditional place as the lone occupant of the aft cabin, however, reflected the central importance of science. Divided in two, the aft cabin housed both Captain George S. Nares and Chief Scientist Charles Wyville Thomson in parallel positions of power. They also shared the 30- × 12-foot fore cabin, which had skylights and a writing table, as their personal sitting room and study.
>
> (Rozwadowski 1996, 415)

These changes in the use of the space of the ship reflects a broader shift in spaces of science and their connection to the "centers of calculation" (Latour in Livingstone and Withers 1999b, 17). For such alterations as those carried out on HMS *Challenger* to be authorized, both the adequacy of the ship as a location of science and the importance of the study of the ocean-space had to be accepted. Certainly, both the space of the ship and the science performed aboard the ship altered the concept of the space-ship and made it more akin to that of the laboratory where science and culture interact in a, in fact, deeply social setting. The importance of the ship emerges as not only a vessel but also as a place of science, a scientific instrument, and a social space.

Science and spaces of science are inextricably linked in a myriad of ways. There are scientific spaces and spaces of science, and the way these are negotiated through standards or the regulation of places bears witness to the need to make sense of this tight entwinement. The ways of negotiating space scientifically highlights the located nature of scientific knowledge. These are ideas that are useful with regard to the making of knowledge about the ocean-space in particular and the Atlantic Ocean in general.

This section considered ways in which knowledge of geographical space can be known from a historical perspective. By initially focusing on geographical imaginations of the tropics and examining how these were produced within cultural discourses, it highlighted the epistemological mechanisms central to making sense of space at a distance. Understanding how knowledge is produced and made to travel without losing integrity is crucial in considering how the ocean-space could become a known geographical space. Indeed, the issues of social class, standardization, and the locality of science are all issues, which will be discussed later. Certainly, as with the tropics, knowledge production about the ocean-space was mediated through a series of cultural factors, which indubitably shaped both its geographical imagination and its geographical understanding. As the ocean-space was mentally discovered and understood, measured and sounded, its geography became known through lines that represented it both as a physical and a cultural space. Olsson (in Pickles 2004, 3) asks, "What is geography if it is not the drawing and interpreting of a line?" Building from the lens of the historical geography of the tropics, this book questions the lines of the geography of the ocean-space through reinterpreting the rhumb lines, social boundaries, and scientific tenets of knowledge production about the ocean-space.

2.2 Historically knowing the ocean-space

The World Ocean is central to all aspects of human life on earth, yet, the World Ocean and individual oceans have traditionally been denied a geography, a history, and a historical geography. Barthes, in fact, asserts that the ocean is a semiological void. He writes that, "in a single day, how many really non-signifying fields do we cross? Very few, sometimes none. Here I am before the sea; it is true that it bears no message" (Barthes in Dening 2004, 13). Thus the ocean-space can easily be construed as a void and certainly there is a significant gap between the ocean as an absolute space, described earlier through a series of statistics, and the ocean-space as a relative space, which the figures sought to illustrate. In this section, I examine how other, parallel ways of considering the ocean-space have contributed to the production of knowledge about it as a geographical space. Before doing so, however, I argue the case for an ocean-centered approach to global history, relocating ocean-space in an area-studies discourse.

2.2.1 History and geography debates in area studies

In a 1999 paper, which stemmed from both interest by the Ford Foundation in ocean histories and the acclaim for their book *The Myth of Continents*, Lewis and Wigen (1999, 161) suggest that maritime studies could be the response to "the crisis in area studies." They write:

> The area-studies model of global scholarship, based on dividing the world into a set number of large, quasi-continental regions, is under assault from a variety of intellectual and institutional forces. New, less rigid models of global scholarship are increasingly being called for by both scholars and funding agencies. One useful alternative (. . .) reframes area studies around ocean and sea basins. Putting maritime interactions at the center of vision brings to light a set of historical regions that have largely remained invisible on the conventional map of the world.
>
> (Lewis and Wigen 1999, 161)

The rendering visible of maritime spaces is key to developing a new theoretical framework that looks beyond otherwise limited models. The need for new ways of examining the world and the desire to seek alternative connections between peoples and places is thus redolent of a wider need to reassess visions of the globe from scholarly, social, historical, and cultural outlooks. In line with postmodernist geography, the turn toward ocean histories and stories of circulation seeks to "emphasize fluidity, contingency, movement, and multiplicity, questioning the rigid spatial frameworks that have limited and constrained geographical imagination" (Lewis and Wigen 1997, 15).

Similarly, Bentley (1999, 215) suggests a move away from the nation-state as a territory by considering

> [the] larger-scale processes that have deeply influenced both the experiences of individual societies and the development of the world as a whole. In combination, mass migrations, campaigns of imperial expansion, cross-cultural trade, biological exchanges, transfers of technology, and cultural exchanges have left quite a mark on the world's past. Adequate study of these processes requires historians to recognize analytical categories much larger than national communities.

Specifically, Bentley (1999, 216) aims to examine global networks not only as transcending national boundaries but, following Lewis and Wigen, conceptualizing the ocean space as an "alternative to national state and the various other terrestrial constructs that scholars, public officials, and the general public have traditionally taken as natural or coherent world regions." However, perhaps more so than landmasses, the delineation of bodies of water is even more ephemeral than territories (see Lewis and Wigen 1999). Bentley (1999, 217) argues that it is necessary to consider and question "the temporal boundaries of large-scale maritime regions, their spatial boundaries, and the relationship of maritime regions both to each other and to the larger world." Linking this with the idea of historically located maritime communities and specific, time-limited trade routes, the maritime space emerges as one whose history is fraught with complex interconnections. In other words, these show that "larger-scale processes and cross-cultural interactions have long been important ingredients in the development of human societies, they are susceptible to analysis and critique as global realities that constitute a part of the longer historical context of contemporary globalization" (Bentley 1999, 222). Certainly, when they are examined and grasped both as concepts and realities, these global maritime relations make way for international, transcontinental, and timeless ways of interpreting the ocean-space.

2.2.2 An ocean-space?

Since the early 1990s, a small but committed academic circle has been questioning the ocean in terms of social and cultural factors that are deeply embedded in human history. For instance, Armitage (2002, 12–13) reminds us that

> there have been Atlantic historians since at least the late nineteenth century; there have also been avowedly Atlantic histories. But only in the last decade or so has Atlantic history emerged as a distinct subfield, or even subdiscipline, within the historical profession.

The aim of Atlantic Ocean historians is to understand the ocean-space better as one that held a place in human development and whose discovery and understanding thereof is at the core of the global system. The routes opened up by increased oceanic knowledge turned the world into the fully networked space of human and technological interaction. This was the first step in making the globalized world and toward globalization. Knowing the ocean-space can be the initial step toward understanding the seeds of today's networked world; however, there are many facets to knowing the ocean in a formal, academic setting. For instance, the localized, past-oriented Atlantic History Seminar at Harvard or the global, future-gazing World Ocean observatory are redolent of this concern in looking at, learning about, and dealing with the world's oceans and the World Ocean with energized interest.[17] Each promotes the understanding of the ocean-space in terms of social, cultural, and intellectual environments. For example, from the historical stance, there is a lot of focus on the "different 'shapes' that historical geographies of the sea might take" or, as others have seen it, the different colors of the Atlantic Ocean (Lambert et al. 2006, 481), referring to different historical strands of the Atlantic Ocean, identifying them symbolically with colors. In effect, what ocean-space offers to geography is the possibility of transcending "the myth of continents" and visualizing the globe as a single, interconnected entity with oceans at its core (Lewis and Wigen 1997). As Lewis and Wigen (1999, 162) write:

> [Geography's] failure to (. . .) sustain a tradition of global-level analysis more generally—has in some ways severed the field from its intellectual roots. By reembracing the global scale and by joining the effort to rethink the ways in which the world is divided and examined, geography may be able to begin reclaiming the central intellectual position it once enjoyed.

From this perspective, the contribution to geography of ocean-centered studies has immense potential in reshaping the geographical debate, bringing to the fore relationships that might otherwise go unnoticed.

Further, the World Ocean observatory advocates a clear human-ocean relationship that would thrive in an ecologically sustainable manner. Soares (in WOO 2016) argues that:

> We need to forge a new ethico-political relationship between humanity and the oceans, a relationship with a political and juridical basis which creates an atmosphere of sharing and solidarity and which provides for a new universalism centered on knowledge of the oceans; a relationship capable of unifying the citizens of the world under one banner, a common, unique and irreplaceable asset: the sea which all the continents share.

In effect, both enterprises seek to have the ocean-space graduate from the non-lieu status and become a cultural entity in itself, complete with a history and a future. As Klein and Mackenthun (2004) write:

> Contemporary debates about globalization, transculturation, contact zones, the multiplicity and non-synchronicity of cultures and histories invite us to regard the ocean as a historical location whose transformative power is not

merely psychological or metaphorical. The sea, whether as the Black Atlantic, the quasi-Arcadian Pacific, or the Mediterranean omphalos, has been the site of radical changes in human lives and national histories.

Knowledge of the ocean-space is tightly entwined with the modern and global world, who's making saw networks of ships explore, measure, map, pillage, pollute, sell, migrate, and discover the ocean-space. All these stories together write the history of the ocean-space. Yet the acknowledgment of the ocean-space as a historical space is not straightforward. Since the discovery of the oneness of the oceans, the ocean-space has been culturally shaped and socially created in a kind of ubiquitous relationship with its opposite, that is land (see Parry 1981).

In fact, geography, while etymologically concerned with the study of earth in its entirety, is, in practice, understood and applied primarily to the world's lands and perhaps some coastal areas.[18] The same is true of history and, by extension, historical geography. Nonetheless, while history has certainly taken place on the ocean (naval battles are a prime example of this fact), I do not consider them to be maritime events or as giving the sea a history. Indeed, these events were part of a larger terrestrial set of events, not independently maritime: in some sense, they were displaced landed-events. Thinking about space has been limited to spaces, which have had long-standing physical interactions with humanity. Thus the majority of scholarly writings in these subjects deal with, approximately, one-third of the entire surface of the earth. Indeed, oceanic spaces have been largely dismissed in these areas and rarely deemed a subject worthy of study, especially if there were no visible economic profit to be made of studying the ocean-space. Indeed, by the middle of the nineteenth century, the main catalyst for scientific and cultural interest in the sea as such was the prospect of financial gain. The laying of the transatlantic submarine cable between Ireland and Newfoundland (which was successful in 1866) provided a reason other than just knowledge for knowledge's sake to sound the deep sea (Rozwadowski 2005, 13–15). Likewise, when *Hunt's Merchants' Magazine* calculated the economic savings that accrued from understanding oceanic currents, Maury's works on sea passages gained momentum and following (Hale 1854, 546–547). Thus the ocean-space has been more often traversed than thought about historically. As Mackenthun and Klein (in Dening 2004, 13) write:

> Like the desert, the ocean has often been read as an empty space, a cultural and historical void, constantly traversed, circumnavigated and fought over, but rarely inscribed other than symbolically by the self-proclaimed agents of civilization.

The stance that the ocean-space is a culturally empty one is a typically accepted one in Western thought and remains difficult to challenge. For instance, Bachelard talks of the "substantive nothingness of water" and W. H. Auden the "barbaric vagueness of the sea" (Bachelard and Auden in Dening 2004, 13–14). Therefore, geography and history never formally examined the ocean-space and its cultural reach, its social structures beyond acknowledging it as a place of transit:

the ocean-space tended to be overlooked or ignored. Because of the fundamentally transitory nature of human-ocean interaction, it is unusual to conceive of the World Ocean as having a relationship with those who sail it. As crossing the ocean is only occasionally an end in itself, human-ocean interactions are seldom an objective but rather tend to be by-products of missions to get from one place to another. This signifies that the relationship with the ocean lasts only as long as the time transit takes a means to an end.

Conversely, the study of the ocean is often understood as being a scientific, oceanographic enterprise in geological time rather than in human time and lacks a philosophical, ideas-centered stance: oceanography is primarily a scientific discipline. Even when the sea is at the center of human-scale historical approaches, as in Braudel's study of the Mediterranean Sea in the Ancient World or indeed this book, the discussion often begins with a geological overview of the area, explaining that the physical features of the basin justify human actions on its coasts (see Braudel 2001, 11–23). Further, because no permanent or even semipermanent (artificial) settlements pepper the ocean-space, and as crossing an ocean leaves no marks on it, it is not a space that can be studied as lands or coasts have been, compared diachronically, relying heavily on past traces. This is in contrast to landscape studies, which rely heavily on the interpretation of material artifacts. While Schein (1997, 662) argues that it is useful to view "the landscape as a palimpsest" and the cultural landscape a "material palimpsest," at sea, any failed exploratory forays are simply swallowed into oceanic depths, often with the records of expeditions into far-off lands and claims to sovereignty: there are few clues to be unearthed in lengthy archaeological digs, no dirty artifacts to dust off, decipher, interpret, and showcase.[19] Maritime archaeology, while helpful, remains sidelined and the historical value of wrecks limited. When the ocean's very nature swallows physical evidence of a human past, drowns its stories, and muffles its players, it forfeits itself as a geographical and historical void. In effect, as far as the spatiality and discourse of the ocean itself is questionable, it is very difficult to envisage an intellectual realm for that phantom-space. Yet there is a growing body of works that seeks to understand retrospectively the ocean-space from a historical geography point of view, as will be seen. For now, though, I will continue to examine ways in which the ocean-space is construed currently as a void historically, geographically, and culturally.

The ocean as a historical void

History has methods; these are usually accepted to shape the discourse that history produces. Historical thought, while intimately concerned with the past, cannot apprehend it entirely, neither through mathematical thought nor theological or scientific thinking (Collingwood 1993, 5). Furthermore, history operates on a number of analytical levels, each of which is particular to specific ends. To analyze the past, Foucault (2004, 9) writes:

> historians have tools that are both built and received: models of economic growth, quantitative analyses of exchanges, demographic development profiles,

studies of the climate and its variations, recognizing sociological constants, descriptions of technical innovations, their spread, their persistence.

As such, history is not a fixed body of works, but is instead a time-and-place specific, culturally bound methodology, which creates a discourse about itself. We can still assert, however, that, in its simplest form, history is systematic in collecting, organizing, analyzing, and labeling historical artifacts into chronological sequences of events. As Foucault (2004, 14) notes, history is

> the work and the implementation of material documents (books, texts, tales, registers, acts, structures, institutions, regulations, techniques, objects, costumes, etc.) which always and everywhere present, in a society, spontaneous or organized forms of remembering.

Artifacts can range from objects of daily life (pots, wristwatches, bags) to more specialized or professional utensils (telescopes, medical equipment, record players), public papers (newspapers, legal documents, public records), and personal objects (diaries, photographs, spectacles). Together, according to the strand of history and subject guidelines, these artifacts from the past help paint a picture of what once was. However, as Starkey (in Schrope 2006, 622) writes, one "can never revisit the past and replicate what went on, [one] can only get a glimpse. That's an inevitable challenge." Yet, with a strict methodology, history has managed to reconstruct and analyze the past through its leftovers.

As previously noted, the artifacts of the ocean are limited. Wrecks, while numerous, are rarely accessible and often decayed. The story that they tell is usually a terrestrial one that happened to end while in transit at sea. This is true of warships and slave ships alike: their histories are essentially land based, if partly played out at sea. Most significantly, there are no traces on the surface of the sea, no marks, or indeed monuments. The human relationship with the sea is transient and ephemeral, physically invisible and mentally absent. This is why the essence of ocean history appears to be marred from the beginning. Likewise, it is, arguably, the nature of the ocean itself that makes it an apparent historical void. Cultural and social factors may be only consequential. In essence, the end product is the same: until one is academically imposed and socially allowed, the ocean has no history.

The human histories of the ocean, or rather the lack of those, are most eloquently expressed by Derek Walcott in his poem "The Sea is History." The poem focuses on the Atlantic Ocean and its close relationship to West Indian history. As Walcott emphatically states, for nearly five hundred years, the history of the West Indies was simultaneously shaped and swallowed by the sea. The poem opens:

> Where are your monuments, your battles, martyrs.
> Where is your tribal memory? Sirs,
> in the gray vault. The Sea. The sea
> has locked them up. The sea is history.
>
> (Walcott 1986, 364)

The poem continues by describing the ways in which Europeans, arriving by sea, proceeded to intrude upon and change native cultures: "brigands who barbecued cattle," "the white sisters clapping / to the waves' progress," "jetting ambassadors" and "the khaki police" (Walcott 1986, 365–367).

Walcott's poem is interesting here for several reasons. As well as actively conceptualizing a terrestrial region's history as closely intertwined with the ocean, it also proffers upon the ocean a historical voice. Before "the men with eyes heavy as anchors / who sank without tombs," the "ocean kept turning blank pages / looking for history" (Walcott 1986, 365). Walcott's sea is "a guardian of a history that has gone unrecorded by traditional Western forms of preserving the past—narrative, museum, monument" (Klein and Mackenthun 2004, 1). Further, the perspective that Walcott is offering by which the history of the West Indies is woven with, rather than parallel to, the history of the ocean is novel. In effect, the poem

> works to restore the sea to the dynamics of the historical process, energizing it for the project of re-imagining, re-writing, and re-membering the past as a complex and polysemic dialogue, a meeting place of different cultures rather than solely the battleground of antagonistic forces.
>
> (Klein and Mackenthun 2004, 1–2)

How, though, is the ocean-space beyond history? Throughout European modern history, the ocean was, more often than not, relegated to the sidelines as no more than a transit zone where peoples and goods were shepherded without ever experiencing it, living it. The survey of European cultural representation conducted earlier supports this view that the ocean-space has been, more often than not, merely a background for novels or artworks that sought to depict other stories or vistas.

The expression coined by Klein and Mackenthun (2004) to describe the process of making sense of the ocean-space historically is "historicizing the ocean." This is an active process, which sets to

> recover in the history of the sea a paradigm that may accommodate curious revisionary accounts—revisionary in the sense of seeing things in new ways, of seeing them differently—of the modern historical experience of transnational contact zones.
>
> (Klein and Mackenthun 2004, 2)

The aims of this process, Klein and Mackenthun (2004, 2) write, are to

> take issue with the cultural myth that the ocean is outside and beyond history, that the interminable, repetitive cycle of the sea obliterates memory and temporality, and that a fully historicized land somehow stands diametrically opposed to an atemporal, "ahistorical" sea.

Further, the historicizing process seeks to expunge the associated idea that the sea is merely a place for lunatics. Indeed, the ocean is, to this day, still perceived and generally represented as a very hostile environment taken on only by a small elite geared up with sophisticated equipment. I am referring here to the high-level sailing and ocean-rowing world, which thrive on setting new records that never fail to capture the imagination of those who remain on firm land. The events are usually widely reported in the media, and the sailors are idealized and idolized: their feats are usually portrayed as a victory against the raging elements. In these events, the image of the ocean as harsh and hostile is exemplified, as it is depicted as a place where the elements are suffered and the conditions endured. The ocean-space, in effect, maintains its image as a "symbol of madness" an "unruly or romantic anti-civilization," thus strengthening Michel Foucault's dichotomy that "reason has a remarkably *landed* quality" while the sea is definitely irrational (see Foucault 2001; Klein and Mackenthun 2004, 2).

Klein and Mackenthun highlight one particular problem at the core of historicizing the ocean: a cultural tendency to contrast the ocean to land as opposites, as the two extremes of a solidly structuralist setting. I have already hinted at this, and will further develop this notion in Chapter 6 with regard to the deep sea. Thus bearing in mind cultural understandings as "a system in interdependent terms in which the value of each term results solely from the simultaneous presence of others," the ocean does indeed stand opposed to land: it is the other when land is the norm (Saussure 1910–1911, 83). Klein and Mackenthun suggest that land and ocean are culturally polarized, making the ocean an ahistorical place by default. Through the process of actively historicizing the ocean-space, Klein and Mackenthun seek to give the ocean a cultural voice. What I suggest here is that historicizing the ocean-space is synonymous with shattering this particular limited and limiting dichotomy, moving away from a linear approach to space and history and moving toward a multilayered, discursive setting. In a world where even wrecks submerged several centuries ago wield political weight, interpreting what the ocean's waves might have to say can certainly be very contentious (Colapinto 2008). By giving the ocean-space a historical voice, I posit that it is feasible to fragment cultural a priori assumptions about history and space.

Historicizing the ocean is an active process that imposes upon the ocean-space a historical plot. Dening (2004, 14) writes:

> We are making theater of the Sea, imbuing the Sea with narrative. We are talking processes. We are talking of the *how* of our history making as much as of the *what*. We are talking tropes and story, fact and fiction, myth and memory, events and agency. Historicizing the Sea is as much a matter of who reads and hears our histories, as who writes and tells them, and what the story is.

In the same way that Walcott interpreted West Indian history in terms of its relationship with the ocean, academic imposition of a reinterpreted history on locales is a complex process. It constitutes a re-visitation of facts and figures that have invariably been shaped by cultural, sociological, and political factors. A voiced-over ocean-space becomes champion in a history in which discursive space is revisited with a poststructuralist concern for "contingencies of identity, the unde-cidability of meaning, and the indeterminacy of the world" (Ryan 1999, 67). A historicized ocean becomes a monument and "the theme and the possibility of a *global history*" disappear and instead emerges a *"general history"* (Foucault 2004, 17). Foucault (2004, 17–18) continues:

> The project of a global history is what tries to reinstate the general form of a civilization, the principle—material or spiritual—of a society, the com-mon meaning to all the events of a period, the law that makes sense of their cohesion,—what we metaphorically call the "face" of an epoch. Such a proj-ect is linked to two or three hypotheses: we suppose that with all the events of a defined spatio-temporal area, with all the phenomena that we have found traces of, we must be able to establish a system of homogenous relations: a network of causality that allows to extract from each analogous relation showing how they symbolize each other, or how they each express the same central nucleus; we suppose, also, that one and same historicity overbears economic structures, social stabilities, the inertia of mentalities, technical habits, political behaviors, and we subject them to the same type of transfor-mation; we posit, finally, that history itself can be articulated in big units—stages or phases—which themselves behold a principle of cohesion.

The ocean-space here provides an exciting platform to challenge nearsighted approaches to history and reveal patterns that go beyond the water's edge. By looking at the ocean differently and shaping a maritime discourse of history and knowledge, historicizing the ocean seeks to "explore from as many angles as pos-sible the global scale of maritime imaginary" (Klein and Mackenthun 2004, 10). This approach, both theoretical and practical, makes an exciting platform for his-torical innovation.

What this historicizing process highlights is that, at the core of the academic, retrospective historicization of the ocean lies the problem of the multifaceted identities and histories the ocean appears to have: the ocean-space is plural.

Dening (2004, 14) writes:

> Not Sea, but seas. A French sea, an American sea, a capitalist sea, an eighteenth-century sea, a Romantic sea, a Viking sea, *moana*—an islanders' sea.

In effect, the ahistorical ocean-space, when challenged, when historicized, emerges as what Linebaugh and Rediker (2000) have dubbed "the Many-Headed Hydra": a polymorphic and complex body that, while essentially ahistorical, in fact hides a plethora of subdivisions (seas) of history.

Thus, while the ocean is essentially a historical void, construed culturally as an ahistorical space, bilaterally opposed to terrestrial history, the making of the historical discourse of the ocean-space is a useful and central step in challeng-ing stratified and unimaginative historical trends. As such, by reading between the waves, a historicized ocean-space provides a global, across-the-seas backdrop to human history. Individual ocean basins that can help reinterpret human interactions across the oceans provide a malleable, exciting, and novel way of dealing with a plethora of cultural, social, or political offshoots of ocean-crossings. This has been true for most of the earth's basins: for example, Braudel (2001) led the way with his milestone study of the Mediterranean Sea, while others followed suit with the Indian Ocean (Hourani 1995; Pearson 2003), the Baltic and North Seas (Kirby and Hinkkanen 2000), and the Pacific Ocean (Blank 1999). Considering especially the Atlantic Ocean, a number of studies have attempted to consider it as a useful way of making sense of the period in European history and to a lesser extent world history, spanning from 1600 to 1900. These have dealt with questions of exchange, conflict, and ideals on either side of the Atlantic Ocean in manners that seek to highlight similarities or differences, parallels or divergences, connections or detachment in stories on either side of the Atlantic Ocean. Studies embrace the transnational, multiethnic and culturally composite nature of the Atlantic Ocean as a liberating characteristic of working with the ocean as a frame of reference. Shunning traditional self-imposed limitations of nations (or continents) as the lowest common denominator of history, Atlantic Ocean history seeks to

> explain relationships that had not been observed before; it allows one to identify commonalities of experience in diverse circumstances; it isolates unique characteristics that become visible only in comparisons and contrasts; and it provides the outlines of a vast culture area distinctive in world history.
>
> (Bailyn 2002, xix)

Armitage (2002, 11) highlights the advantages of Atlantic history thus:

> The attraction of Atlantic history lies, in part, in nature: after all, is not an ocean a natural fact? The Atlantic might seem to be one of the few historical categories that has an inbuilt geography, unlike the history of nation-states with their shifting borders and imperfect overlaps between political allegiances and geographical boundaries.
>
> (Armitage 2002, 11)

The Atlantic Ocean thus provides natural geographical and methodological guidelines to frame, if not direct, studies of a vast zone with much intertwined history, in an exciting, globalized way that transcends the human and ephemeral borders.

Further, Armitage argues, the concept of Atlantic history is useful to conceptualize, in a strict sense, European history. The Atlantic Ocean, he argues, "was a European invention" (Armitage 2002, 12). He writes:

> [The Atlantic Ocean] was the product of successive waves of navigation, exploration, settlement, administration, and imagination. It did not spring fully formed into European consciousness any more than 'America' did, though it could certainly be found on maps—and hence in minds—two centuries before the full extent and outline of the Americas would be. It was

a European invention not because Europeans were its only denizens, but because Europeans were the first to connect its four sides into a single entity, both as a system and as the representation of a discrete natural feature.

(Armitage 2002, 12)

The Eurocentric aspect of Atlantic history is essential. This is explained, in part, by the fact that Atlantic history and geopolitics are closely linked to the story of European imperialism. Indeed, as Canny (2001, 399) notes, most scholars of Atlantic history are concerned with European aspects of "exploration, discovery, voluntary and involuntary migration, overseas settlement and trade." The mastery of the sea and the desire to control its trading routes were at the core of the imperial enterprise, to which all the aforementioned are tightly linked. Returning to Steinberg (2009, 481), "the idea of an intervening, marine space between the core and the periphery was crucial to the discursive and material workings of the era's European empires." Certainly, as aforesaid, imperial ambitions were one significant aspect of the exploration of the ocean-space. But this is of interest here only in relation to a wider perspective of knowing the ocean-space. For instance, in the case of the British Empire, Harland-Jacobs (1999, 237) has argued that "Britain used the Atlantic as a primary artery for extending its influence overseas." She continues:

The development of the Caribbean plantation economy, the pursuit of the slave trade, the growth of settlement colonies in North America, the establishment of an informal empire in South America, and the migrations of the Scots, Irish, Welsh, and English combined to make the Atlantic a linchpin of the empire for three centuries.

(Harland-Jacobs 1999, 237)

As the nineteenth century saw unprecedented growth and the height of most of the world's modern empires, the world's oceans were the "international and interconnective" links between home and the colonies (Harland-Jacobs 1999, 238). Indeed, "the empires of this period were simultaneously national and global, as Britain, France, and the Netherlands (and later Germany, Belgium, the United States, and Japan) extended their polities, economies, and cultures across the world's ocean" (Harland-Jacobs 1999, 238). Therefore, as the ocean is at the center of imperialism, so is imperialism a core component of the history of the Atlantic Ocean in European thought, strongly influencing studies that look at, say, racial identities, slave-trade economies, or sugar plantations.

However, there are countless ways of doing Atlantic history, and they are not all directly linked to empire and colonization. These include, as surveyed by Harland-Jacobs (1999, 237),

the sugar nexus between planters and slaves in the colonies and bankers, entrepreneurs, and consumers in the metropole, the contours of the Atlantic economy, cultural flows and the transformation of the early American landscape, and transatlantic migration.

Yet the most prosperous subcategory of Atlantic history appears to be those litera-tures derived from specific interest groups that seek to make sense of the Atlantic world in terms of one particular physical or cultural feature. The most prominent of these strands is perhaps Gilroy's concept of the Black Atlantic, which consid-ers the problematics of being both "European and black" (Gilroy 1993, 1). Gilroy focuses his interest on various aspects of the human and cultural consequences of the Atlantic Ocean as a transit zone. The transitory nature of the relationship with the sea is both consequential to its nature and causal to its relegation out of the general culture. Therefore, the ocean-space as one of transit emerges as a focal point in this study, an essential and useful way of making sense of the sea. The Black Atlantic, however, unlike the relationships it is built upon, is not transient and studying its trends provides an exciting and telling new outlook on European, African, and American histories through the Atlantic Ocean.

Others have also worked on the colors of the Atlantic Ocean. Lambert et al. (2006, 481) discuss two other colors from this academic rainbow:

> 'White' is used to denote traditional interpretative models, so colored because of the focus on European (especially British) migration and trans-oceanic cultural transference. In this schema, the 'white Atlantic' is held to be the 'inland ocean of Western civilization'. (. . .) Such perspectives [as the white and black Atlantics] have done much to foster a third approach, that of the 'red Atlantic', which is exemplified by Peter Linebaugh and Marcus Redik-er's *The Many-Headed Hydra*. Exploring the formation of a multinational, multi-ethnic, multicultural Atlantic proletariat, it probably owes as much to the long tradition of 'black Atlantic' historiography as it does to the writings of Marx. The former has also inspired more self-conscious and critical treat-ments of the 'white Atlantic'. Broadly, then, a contrast can be drawn between integrative and developmental historiographies of exploration, colonization, empire-building and the emergence of new states on the one hand ('white'), and circulatory, transformative and postcolonial perspectives on the other (variously termed 'black', 'red' or more critical 'white' approaches).

Further, alongside the White Atlantic, there is also an offshoot of green Atlantic studies. These look at the Irish place and role in the transatlantic (Whelan 2004). Overall then, it emerges that various historiographic impulses paint the Atlantic Ocean in different colors, giving it many hues.

Another way into the study of the history of the Atlantic Ocean is through spe-cific places: this is what Armitage (2002, 15) calls "cis-Atlantic history," which "studies particular places as unique locations within an Atlantic world and seeks to define that uniqueness as the result of an interaction between local particularity and a wider web of connections (and comparisons)" (Armitage 2002, 21).

This approach looks at specific locales and their relationship with the wider Atlantic world. These locales can be very small (an island, a city, an estuary) or much bigger (a country, a coastal region). For instance, Chaunu and Chaunu (1955–59) look at the special relationship between the city of Seville and the

Atlantic Ocean. Sacks (1991) studies Bristol's change from a wine-oriented economy to one centered on Atlantic staples. Rudel (2002) offers a study of the history of the Azores, highlighting the archipelago's central role in the establishment of an Atlantic world. Covering larger zones, Thornton (1998) looks at the role of Africa in the making of the Atlantic world, Francis (2006) concentrates on Iberia and the Americas, and Evans (1996) works on the Irish influences and Atlantic heritage. These cis-Atlantic studies seek to make sense of zones in terms of their relationship with the Atlantic world as much as their role in shaping the Atlantic world. Knowing these zones as cis-Atlantic provides new and exciting perspectives for examining areas, big and small, that share a common Atlantic history and whose respective histories were dramatically altered by the Atlantic world.

These different ways of historicizing the ocean-space demonstrate how the ocean-space can indeed occupy a place in human history that goes beyond being a setting. Instead, it becomes imbued with a sense of place that has changed throughout the years and has actively shaped happenings on, across, or around it.

The ocean as a geographical void

From a strictly theoretical point of view, it can be argued that the ocean, rather than being a space, appears to have traversed European geographical thought as a non-space, an invisible, specter-like scene where nothing of interest to mankind ever happened. For instance, ships' logs were useful only for captains and businessmen but held little sociocultural value. These were indeed significant only within the tightly monitored networks of trust, as discussed earlier, and bore little connection beyond these networks. While interior seas had been essential for short-distance trade—the Mediterranean Sea and the channel especially for Europe—the World Ocean did not exist beyond these useful, usable, and known limits. Ships and goods traveled along coastal areas, over significant distances, but land was rarely out of sight. Even during the Age of Discovery, characterized by its audacious voyaging and pushing the limits of knowledge, what interested European explorers was not the ocean itself, but the limits of the ocean, what lay beyond the ocean, how to make use of the ocean for commercial, religious, or imperial ends. The ocean-space was just the path to other spaces, not the space itself. For example, when the Atlantic Ocean was eventually discovered as we know it today, what mattered was not the Atlantic Ocean in and for itself but the potential it held as a gateway to other, new places, such as, for instance, the tropics.

On these grounds, the ocean-space can be considered a non-lieu, or non-place. The term "non-lieu" literally translates as non-place, but it also has legal connotations: a legal non-lieu is when a case is deemed not worthy of being taken to court. Non-places are, following Augé, those

> spaces associated with transit and communication, designed to be passed through rather than appropriated, and retaining little or no trace of our passage as we negotiate them. [Non-lieux] are the product of technological advances which have vastly increased mobility, both local and global, and

thus produced massive urbanization and migration, are devoid of meaning, identity or community. On the contrary, they constitute "un monde [. . .] promis à l'individualité solitaire, au passage provisoire et à l'ephemère."

(O'Beirne 2006, 38)[20]

Opposed to places, which are culturally distinctive, imbued with social practices, marked by human activity, non-places are anonymous, devoid of cultural significance, interchangeable, silent. Augé (1995, 77–78) writes:

> If place can be defined as relational, historical and concerned with identity, then a space which cannot be defined as relational, or historical, or concerned with identity will be a non-place.

Augé's own examples of non-places include modes of transport, their associated transit zones, as well as supermarkets, shopping malls, and other faceless depositories of goods. These are all modern spaces and spaces of modernity that resulted with increased global networks of peoples and goods. He writes:

> Place and non-place are rather like opposed polarities: the first is never completely erased, the second never totally completed; they are like palimpsests on which the scrambled game of identity and relations is ceaselessly rewritten. But non-places are the real measure of our time; one that could be quantified—with the aid of a few conversions between area, volume and distance—by totaling all the air, rail and motorway routes, the mobile cabins called 'means of transport' (aircraft, trains and road vehicles), the airports and railway stations, hotel chains, leisure parks, large retail outlets, and finally the complex skein of cable and wireless networks that mobilize extraterrestrial space for the purposes of a communication so peculiar that it often puts the individual in contact only with another image of himself.

(Augé 1995, 79)

In this respect, as indeed a transit zone for exploration, the World Ocean emerges as a non-place whose significance is measured not in itself but for where it could lead. As a non-space, the ocean-space is first and foremost transoceanic. In practice, the reality of oceans existed inasmuch as they were transit zones. Despite being increasingly crossed, defined, measured, mapped and, ultimately, known, the ocean-space long remained a place not physical enough to have a geography and, as shown earlier, a history. Instead, its existence was only in imaginations as a non-place twice removed from sociocultural goings-on, not visible enough to be studied or even acknowledged.

From a practical standpoint, the ocean's nonexistence in geographical thought is visible in several different ways, as I will now demonstrate. These examples, while very localized, seek to incorporate the frame of mind of general attitudes toward the ocean-space, namely that it has no space or place.

Conventionally, and as a means of convenience, when nations have associated islands, these are drawn, on the nation's map, in an inset. The island is not necessarily represented to the same scale as the rest of the map, and the position of the inset itself bears no relation to the actual relative position of the island to the mainland. For instance, on a political map of Spain, the Canary Islands are usually inset below the Balearic Islands instead of 250 kilometers off the coast of Morocco. Further, and more to the point here, the sea in which the island lies is erased, made invisible, ignored. The value of the ocean-space itself is irrelevant. One example of this Atlantic Ocean wipeout is the place of the island of Rockall, in the Outer Hebrides, on maps of the United Kingdom (Figure 2.3). Rockall is a rocky islet situated 280 miles west of the Isle of Skye. At its most, it is twenty-seven meters wide and, due to the fact that it "cannot sustain human habitation or economic life of its own," its ownership provides no economic benefits to the country: Rockall yields no territorial waters (Charney in MacDonald 2006, 643). Yet Rockall has associated value: for instance, it was at the center of a battle over "perceptual fields" during the Cold War (MacDonald 2006, 635). Today, however, it only comes with unresolved issues pertaining to the right to exploit the continental shelf for any resources underneath the ocean floor (Symmons 1979, 101–102; Symmons 1986). Here the importance of this uninhabitable rock is that, while it is part of the United Kingdom, it is officially outside the British national grid reference system (OS 2008, 26). Therefore, it is officially represented in an inset with no gridlines. Yet Rockall is as much part of the United Kingdom as is Nelson's column in London. Rockall's cartographical isolation is redolent of a tendency to erase the ocean-space from geographical representations because it takes up too much paper-space.[21] Further, due to the small size of the islet, Rockall has to be represented at a different scale than the rest of Britain. Otherwise, it would hardly appear as a dot. Therefore, the official maps of Britain are neither coherent, nor entirely geographically comprehensive.

What this example illustrates is the secondary nature of the ocean, not only in geographical thought but also in cartographic representations. Omitting to represent the ocean-space on maps can be construed as relegating the oceans as unimportant. In the case of mapping the Pacific Ocean, Cosgrove reminds us how it came almost as an afterthought. He writes:

> The Pacific as a bounded geographical—or, better, oceanic—space emerged relatively slowly, from the gradual extension of coastline charting and accurate fixing of island locations, a process that lasted until remarkably recently with the final delineation of Antarctica. Already split and pushed to the margins of the Western world image by the seventeenth-century Dutch innovation of the double-hemisphere map, Pacific space, even in the mid-nineteenth century, presented cartographic problems, whose (partial) resolution during subsequent decades shifted the narrative away from discovery and towards global geopolitics.
>
> (Cosgrove 2008, 185)

Figure 2.3 Winter wave breaking over the island of Rockall, 11 March 1943. Photographed from an RAF Coastal Command aircraft from RAF Aldergrove during World War II.

Source: https://wikipedia.org/wiki/Rockall#/File:Rockall_wave_March_1943.jps

Yet, while they are demographically and politically lightweight, there are associated powers that come from having ocean-space. For instance, the Exclusive Economic Zone (EEZ) gives a nation rights over 200 nautical miles of sea from its coast, and nations have rights on the continental shelf up to 350 nautical miles from their coast. EEZs are considerable assets, covering all economical and commercial aspects of fishing and mining within their limits: they are the extension of the terrestrial country.[22] However, conventionally, they neither appear on political maps nor are they thought of as part of countries (Steinberg 2009, 487–488).

Another way of considering the imposed insignificance of the ocean-space is connected with the legal aspect of owning the sea. I examine this issue further at the end of this chapter, but at the moment, I will focus on the matter of national waters. According to international law, sovereign states with shoreline are allowed to claim control over a certain amount of sea that is directly adjacent to its shores. These are territorial waters. The laws governing the exact amount of sea allowed are complex and have changed many times in the past, but the internationally accepted distance is twelve nautical miles, unless this clashes with another country's claims.[23] For instance, in the case of the English Channel/la Manche, the boundary between France and the United Kingdom runs in the middle of it, except where the Channel Islands give the British the right to claim sea up to almost the entrance of St. Malo.[24] Indeed, a judgment passed by the International Court of Justice in 1953 confirmed sovereignty of the Minquiers to the British crown, and although the rocks have been twice invaded by the kingdom of Patagonia, this means that the central portion of the Channel is mostly British (Deniau 2002, 357).[25] In all events, this aspect of maritime law means that areas under national sovereignty are much greater than the earth's terrestrial surface (ICJ-CIJ 1953a and ICJ-CIJ 1953b). Looking at a globe in this light, then, the Federated States of Micronesia, in the Pacific Ocean, covered a distance greater than that from Paris to Moscow. Yet rarely is it thought of in that way. Micronesia, to all intents and purposes, is hardly more than a collection of 607 small islands, not a political entity comparable in size to Europe. This is because the ocean-space has been taken away from it, reducing it to landmass when the core of it is sea.

This brings me to the matter of the seabed and its ownership. Recently, Russia laid claim to the seabed directly under the North Pole. Later, Australia extended its own claims to water around Tasmania and, more recently, Britain sought to extend its claim to waters surrounding Ascension Island to two hundred miles (BBC News 2007, 2008a, 2008b). While owning waters around an island or planting a flag under a pole may seem a futile gesture, the political value of these acts must not be underestimated. The ownership of the sea stands for more than the water it represents: it includes the right to fish in these waters, the right to the seabed and its minerals, and the right to what gas or oil may be contained in the seabed. Thus, while the initial interest may appear to be in the ocean itself, what these political bodies are, in reality, concerned with are the assets attached to the ocean floor and all the fish and marine life that live in it. Once more, the physicality of the sea is overlooked, as the water itself is devoid of valuable substances.

These three examples highlight the geographical and geopolitical emptiness of the ocean in Western terms. From the blatant subtracting of it on maps, to the simple omission of it in minds, and the forgetting of it in terms of resources, the ocean-space lacks geographical substance. In minds as well as on maps, the ocean-space remains a geographically challenging space, but through a process of what can be called "geographicalizing" (a process that is tightly entwined with the processes of historicizing discussed earlier), it becomes possible to construct a geographical ocean-space.

The geography of the ocean-space and the process of geographicalizing it is manifold: while on the one hand it comprises all aspects of measuring the ocean-space and cartographically representing it, it also needs to include the mental process of discovery of vast spaces. Importantly, these processes that are at the core of knowing the ocean-space are essentially artificial academic ventures: in a Western mind, the geography of the ocean-space exists only inasmuch as one was instituted upon it. From a Western perspective, having a geographical ocean(-space) includes scientific (absolute) and cultural (relative) ways of knowing the World Ocean. The former of these is synonymous to having geographical coordinates that accurately describe the limits and properties of the World Ocean. If the Scientific Revolution urged a new way of knowing things and making facts about nature, it also quietly imposed a system that equated lack of facts to nonexistence. The Enlightenment was, in this way, a totalitarian enterprise that left no room for emotion, seeking instead to rationalize and mathematicize everything. Therefore, as the Age of Discovery was turning space into numbers, it squeezed out of consciousness everything that did not fit within the framework of newfound reason. As Adorno and Horkheimer (2002, xvii) write, "the flood of precise information and brand-new amusements make people smarter and more stupid at once." Within the context of the Age of Discovery, and through the Enlightenment's program of "disenchantment of the world," the ocean-space became known through numbers and measurements (Adorno and Horkheimer 2002, 1). Myths about it were indeed dismissed, as the ocean-space was gradually mapped and its nature was objectified. Yet, while ignorance about the ocean-space was turned into data, so was its geographical imagination. Once quantified, the World Ocean was demystified. Indeed,

> Enlightenment, understood in the widest sense as the advance of thought, has always aimed at liberating human beings from fear and installing them as masters. (. . .) Enlightenment's program was the disenchantment of the world. It wanted to dispel myths, to overthrow fantasy with knowledge.
>
> (Adorno and Horkheimer 2002, 1)

For instance, in Europe, the Atlantic Ocean became important and interesting only for the facts that could be extracted from it and, especially, those that could be observed from its shores. In eighteenth-century Europe, the seashore became the place to go and observe the ancientness of the earth and attempt to visualize geological time (Corbin 1994, 97–120). As such, the Atlantic Ocean was useful only on the periphery.

Geographicalizing the World Ocean is thus a motion toward doing two things that a terrestrial geography approach has failed to do with this particular ocean-space. First, it insists on replacing the cultural aspect of the ocean-space within an otherwise clinically measured and surveyed space. Second, it is to move away from the coastline as the basis for contact with and creation of the cultural ocean-space. Again, this is to locate the place of the World Ocean within a larger geographical framework that goes beyond European coasts and places it in a global and networked world. With regards to the Atlantic Ocean, as Ogborn (2005, 379) writes:

> Atlantic geographies require perspectives that can encompass spatialities that include but exceed those of the local or of the nation-state, and historical periodizations that are attentive to the long term and to dynamic, circum-navigatory flows that disrupt notions of progress and development. Studying Atlantic geographies means opening perspectives on a social, cultural and spatial arena which has to be considered as a complex unity, but which also needs to be differentiated in terms of the many journeys, peoples, ways of life, ideas and materials that have been brought together in both violent and productive ways in the Atlantic world.

This process will move geography away from being solely about new lands discovered and the possibilities (physical, economical, commercial, etc.) of these new lands to include the space in between. This geography has yet to be imbued with meaning, and only then will we be able to speak of World Ocean geographies.

2.3 Historical geographies of the world ocean

An entry dated late-1600s in Alexander Olivier Exquemelin's diary reads thus:

> "[The green sea turtles'] eggs are found in such prodigious quantities along the sandy shores of those countries [in the Caribbean and along the South American coast] that, were they not frequently destroyed by birds, the sea would infinitely abound with tortoises . . . certain it is that many times the ships, having lost their latitude through the darkness of the weather, have steered their course only by the noise of the tortoise swimming that way, and have arrived at these isles.
>
> (Exquemelin in Schrope 2006, 622)

From this passage, the assumption could easily be made that Exquemelin was a biologist-explorer traveling in the area, recording his observations in order to compile a natural history of the zone upon his return to Europe. Exquemelin would not have necessarily been particularly original in this enterprise, as the period inspired many aspiring explorers to spend time studying the fauna and flora of the newly discovered region. In reality, though, Exquemelin was a French-born pirate and the excerpt is from his book, *Buccaneers of America*, which is a detailed account of all aspects of pirate life in the West Indies in the late-seventeenth century. Thus it surprises as a viable source of historical information and knowledge about the ocean and its ecosystem. Two other examples of uses of unexpected historical artifacts used to know the ocean are cookbooks as useful sources to gauge fish populations and surveyors' charts, which provide reliable secondary data on such things as coral reefs, which would not usually appear on maps (Schrope 2006).

The tools through which we know the ocean historically are limited: no artifacts or settlements are here to guide us. Therefore, thinking about historical geographies of the Atlantic Ocean requires creativity and originality in terms of source materials. In particular, one program that uses original ways of exploring the oceans' past is the History of Marine Animal Populations (HMAP 2011). The aims of this project (which is itself contained within a larger undertaking called the Census of Marine Life that wants to catalogue everything that lives in the World Ocean) seeks to correct "historical myopia" and what has been called the "shifting baseline syndrome" (Smith and Pauly in Schrope 2006, 622). These problems stem from both the absence of long-term records of marine populations and the difficulty of retrospectively compiling reliable information on past marine populations, hence the turn to more original sources, such as personal journals.

In one branch of the study, McClenachan and Jackson examined the writings of William Dampier, another buccaneer, to find hints about turtle populations in the Caribbean in the late sixteenth century (see Schrope 2006, 622). They also sought to retrace the evolution of coral reefs around the Florida Keys, "as well as the population history for 14 important species including mangrove trees, grouper fish and the now-extinct Caribbean monk seals" (Schrope 2006, 622). In such research, passages like the following one extracted from the observations of a surveyor in 1796 are, again, useful sources of information. Indeed, of the mangroves in the Keys, George Gauld (in Schrope 2006, 623) wrote that "among the roots of the mangroves and about every old log or piece of rotten work, there are such quantities of the largest crayfish that a boat may be loaded with them in a few hours." Schrope tells us that this would be impossible today, due to a severe decline in crayfish stocks. In other places, other kinds of sources also show changes in fish populations. In the Wadden Sea region, sturgeon was one affordable species among the popular classes, yet,

> by the fourteenth century they were so rare that they were only served at the king's table. Later, French cookbooks held clues to the species' drop in numbers, such as instructions on how to prepare veal as a substitute in recipes calling for unavailable sturgeon.
>
> (Schrope 2006, 623–624)

This steady decline in numbers was caused by human interference, mainly overfishing and the construction of dams. On the other side of the Atlantic Ocean, a similar trend was visible in cod populations. When Atlantic Ocean cod fisheries collapsed in the 1990s, Emmanuel Altham, a fisherman in the area in the 1620s would have been amazed. Indeed, in 1623, he had written in his diary that:

in one hour we got 100 great cod . . . and if we would have but stayed after the fog broke up, we might quickly have loaded our ship . . . I think we got 1000 in all . . . When we had nothing to do my people took great delight in catching them, although we threw them away again.

(Altham in Schrope 2006, 624)

Similarly, Richard Whitbourne (in Rose 2007, 205) wrote, around the same time that "the island of New-founde-land is large, temperate, fruitefull (*sic*) . . . the Seas are so rich, as the are able to advance a great Trade of fishing; which, with Gods (*sic*) blessing will become very serviceable to the Navie." Yet, in the early-1990s, fisheries in the North Atlantic were dismissing staff after "the 'perfect' demise of the northern cod" (Rose 2007, 205). It was perfect because it was "created by so rare a combination of factors that it could not possibly have been worse" (Rose 2007, 469). What these new research projects do is look at the world or, in fact, the ocean, through new lenses. The Atlantic Ocean is historically rediscovered in terms of the ecosystems that it supports.

While such studies are exciting and the outputs they produce novel, they are still one step away from historical geographies of the World Ocean. These historical geographies may not exist, but certainly fish stocks constitute a significant part of human understandings of the ocean, and these new ways of considering it highlights the human relationship with the bounty of the sea and shows us tangible correlation between the inhabitants of the coastal area and the ocean itself.

Another useful way of making sense of this interconnection between the ocean-space and its actors is by looking at the manner in which the oceanic areas were made sense of in a legal sense. In fact, with the twin discovery of the Atlantic and Pacific Oceans (discussed in the following chapter), questions about how to govern these new spaces were raised. From a European point of view, oceanic expansion "reconfigured international relations, obliging expansionist powers to define the legal and diplomatic implications of interstate conflict in the extraterritorial arena of the high seas" (Mancke 1999, 225). Ocean-space, Mancke (1999, 225) argues, is a space where empires could be played out in manners drastically different than on land while reassessing the agency of indigenous peoples in the early history of the modern world and reconfiguring international relations. If the exploration of the seas lying west and south of Europe began mainly unpoliticized, by the end of the seventeenth century, "the control of oceanic space had become not just a commercial question but part of the construction of power in the European state system" (Mancke 1999, 233). Even today, when there are still as many as sixteen colonies speckling the oceans, these territories still represent strongholds of political power and military purposes, emphasizing the importance of the ocean-space as one of power and politics.[26] The result is that, Mancke (1999, 226) writes:

The politicization and militarizations of oceanic space, as much as its global-ization, distinguished European oceanic expansions from that of other sea-faring peoples. (. . .) [When] Europeans began their long-distance maritime ventures, trade and colonizations were old processes in the Indian and Pacific oceans. But Europeans' transoceanic political claims and their attempts to control and regulate access to the high seas were new phenomena.

The discovery of the Atlantic Ocean, for example, was accompanied by a rush to at least make legal sense of the new space discovered.

Steinberg (1999b, 254) suggests that marine governance trends can be divided into two: "alternating currents for and against division and territorial enclosure," he explains:

> On the one hand, [historians of marine governance] note, events and proclamations such as Hugo Groitus' 1608 *The Freedom of the Seas*, the nineteenth-century "free-seas" policy imposed under Pax Britannica, and the nonterritorial self-regulation practiced by the maritime-transport industry in the twentieth century represent attempts at constructing the ocean as a friction-free void wherein nascent colonial empires and enterprising merchants could establish lines of connection with far-flung terrestrial territories, production sites, markets. On the other hand, as interaction with the ocean has intensified over time, the ocean itself has come to be perceived as a space of resources, whether the resource is that of connection or something more material, such as fish or minerals.
>
> (Steinberg 1999b, 254)

These ways of dealing with the ocean-space, while inherently at different ends of the spectrum, highlight the contradictory desire to both control the ocean-space by enclosing it within political lines and the eagerness to construct it as a space of transit and flow that would be frictionless and internationally harmonious.

One of the most famous instances of prorating the ocean-space took place between 1493 and 1494, dividing the world in two parts down a line of longitude that was haphazardly placed 100 and then 370 leagues west of the Cape Verde Islands. In 1493, the "Inter Caetera" papal bull was issued following Christopher Columbus's return from his first journey. This bull is addressed to the king and queen of Spain, remaining silent regarding Portugal. The object of the bull is to ensure that Christian rule be expanded on whatever lands "discovered or to be discovered" (Alexander VI 1493). Of the inhabitants of these lands, the bull explains:

> envoys are of the opinion [that they] believe in one God, the Creator in heaven, and seems sufficiently disposed to embrace the Catholic faith and be trained in good morals. And it is hoped that, were they instructed, the name of the Savior, our Lord Jesus Christ, would easily be introduced into the said countries and islands.
>
> (Alexander VI 1493)

It continues:

> [As] becomes Catholic kings and princes, after earnest consideration of all matters, especially of the rise and spread of Catholic faith, as was the fashion

of your ancestors, kings of renowned memory, you have purposed with the favor of divine clemency to bring under your sway the said mainlands and islands with their residents and inhabitants and to bring them to the Catholic faith. Hence, heartily commending in the Lord this your holy and praiseworthy purpose, and desirous that it be duly accomplished, and that the name of our Savior be carried into those regions, we exhort you very earnestly in the Lord and by your reception of holy baptism, whereby you are bound to our apostolic commands, and by the bowels of the mercy of our Lord Jesus Christ, enjoy strictly, that inasmuch as with eager zeal for the true faith you design to equip and despatch this expedition, you purpose also, as is your duty, to lead the peoples dwelling in those islands and countries to embrace the Christian religion; nor at any time let dangers or hardships deter you therefrom, with the stout hope and trust in your hearts that Almighty God will further your undertakings.

(Alexander VI 1493)

The "Inter Caetera" bull then, rather than dividing sea, is in fact giving land to one specified party. Furthermore, these lands are allocated with a particular purpose in mind: to make Christian those unidentified lands and undiscovered populations before anybody else could claim them.

Portugal's exclusion from this allocation of lands afar caused some political tension in Europe (Spain and Portugal were the only two European powers strong enough to consider expansion at this stage). By 1494, enmity had mounted, and a conference was called in Tordesillas to resolve the issues arising between Spain and Portugal. Having first drawn the Tordesillas line 370 leagues west of the Cape Verde Islands, the treaty agreed that all lands to the east of the line would be Portuguese and all those to the west would remain under Spanish control. Thus, in the space of a single conference, the world was divided up and control of lands assigned.

Yet, while the Tordesillas line divided the land, so it did too the Atlantic Ocean.

However, neither the bull nor the treaty mentioned the ocean-space. Steinberg (1999b, 257) clarifies that:

> Spain's and Portugal's claims to exclusive right should not be viewed as claims to possession of the sea. Rather, the two countries' claims implied that the sea had been divided into 'spheres of influence' in which Spain and Portugal were granted rights of stewardship. (. . .) The stewarding entity is presumed to have a right to exert control both over the resource or space being stewarded and over others who might wish to use the stewarded resource in a contrary manner.

The Atlantic ocean-space is here stewarded, not owned, and certainly not equal to lands and territories.

While on the one hand, control over the ocean-space is beneficial in terms of where it can lead, even when space is distributed politically, it fails to be acknowledged as a space, neither political nor commercial. From this perspective, the geography of the Atlantic Ocean fails to overcome the politics that are unable to see it. The idea of lines demarcating vast expanses of water that are twice removed from their governing bodies is geographically questionable. The lines are hardly more than political constructions that exist, if not in practice, solely politically and with few ramifications affecting the ocean-space itself. This invites us to think about the greater relationship between society and space as played out on vast expanses that are distributed while hardly being mapped and often still only exist in popular and sociocultural imaginations. Here the geography of the Atlantic Ocean emerges as a by-product of other, land-based political movements, not an end in itself.

2.4 Conclusions

In spite of methodological and cultural challenges that facilitate ignoring the ocean-space culturally and geographically, this chapter has begun to tease out ways in which the alleged emptiness of the ocean-space in terms of historical and geographical standpoints can be filled by reaffirming the World Ocean as a space of history and geography. It can be culturally unearthed (or unsea-n) by locating the ocean-space against the backdrop of contemporary debates in geography and area studies. By thinking about the ocean-space in terms of historical and cultural contexts, as well as within a changing geographical framework, the World Ocean becomes a useful way to trace the intellectual and cultural changes related to the ocean-space. Considering the ocean-space historically and geographically, through cultural representations and shifting legal structures, provided a useful perspective on how the ocean-space has been understood as a geographical and physical space. Indeed, in the words of Mary Oliver, "the sea / isn't a place" but, as Steinberg writes, "a space of contradictions and alternatives, of images and laws, of labor and dreams" (Oliver in Steinberg 2001, 210). By bringing to the forefront the cultural importance of the World Ocean and seeking to think about the ocean-space historically and geographically, the physical and intellectual paths of the discovery of the World Ocean become clearer.

Notes

1 This is equal to roughly 170 times the area of metropolitan France.
2 This is approximately equivalent to the volume of a cube whose side is 678 kilometers or a sphere whose radius is 420 kilometers.
3 The average spreading rate for the Atlantic Ocean is about 2.5 centimeters per year (USGS 2014). Comparatively, a human fingernail grows at the average speed of 2.5 centimeters a year, meaning that the Atlantic Ocean widens by approximately the same distance as a fingernail grows (Dawber 1969).
4 Taking a bathtub's volume as about 180 liters, 1 Sv is equivalent to emptying 5.555 million bathtubs in one second. Therefore, the flow of the Gulf Stream is equivalent to emptying between 166,650,000 and 833,250,000 bathtubs per second.
5 For the time being, the term "non-space" is used to refer to the cultural and intellectual nonexistence of the Atlantic Ocean as a geographical space. I discuss the ocean-space as a non-space later.

6 See Bravo 2006; Driver and Martins 2002 and 2006; Duncan 2000; Harland-Jacobs 1999; Lorimer 2003; MacDonald 2006; Naylor 2000a, 2005b, 2006; Ogborn 1998, 2000, 2004; Powell 2008; Ryan 2006 for examples of this kind of historical geography.

7 When Verne was writing, it was believed that Antarctica was floating ice, like the Arctic. It was not confirmed to be a continent until later.

8 The *oikoumene* is "the habitable earth" (Cosgrove 2008, 105).

9 However, not everyone was readily inclined to trust these "simple people." For example, as I discuss in Chapter 4, some English captains refused to heed advice from "simple American fishermen" (Franklin 1806, 186). However, the English refusal might have been caused both by the fishermen's simplicity and their Americanness.

10 See Anderson (2006) for more on "imagined communities."

11 See Cook and Lux (1998) for more on the implications of weaker ties.

12 See Cook (2006) for more on Dalrymple, his role at the British Admiralty, and his contribution to hydrography.

13 For more on Latour's approach to the question of localizing science and the space of the laboratory, see Latour (1986 and 1987).

14 With the advent of satellite navigation systems in the late-twentieth century, a similar problem emerged. Indeed, each country wanted, again, its own standard base, and therefore maps were once more drawn to different standards. Today, several coordinate systems are in use, including WGS84 (which is very widely used).

15 Another comparable example of targeted standardization has to do with the adoption of a standard time. In Britain, the initial push for standardized time came from the post office as it began running all its coaches according to a Greenwich Mean Time (Zerubavel 1982, 6). This was, according to Zerubavel (1982, 6) "the first attempt in history to synchronize different communities with one another." In the United States, however, Bartky (1989, 25) writes:

> standard time was not adopted primarily to bring order to the chaos of railroad time-tables. Indeed, the railroads did not need standard time for their operations. Rather, in the 1870s scientific pursuits requiring simultaneous observations from scattered points became important, and those needs led to proposals for federal action in the early 1880s. In response to these pressures from scientists, railroad superintendents and managers implemented a standard time system on November 18, 1883, a system tailored to their companies' train schedules.

> However, Schegloff (in Zerubavel 1982, 1) notes that there is still such a thing as "temporal formulations" that fall outside the standard. One example of this is found in Kurt Vonnegut's novel *Cat's Cradle* in which a character says, "When I was a younger man—two wives ago, 250,000 cigarettes ago, 3,000 quarts of booze ago" (Vonnegut in Zerubavel 1982, 1). This is akin to the Bergsonian concept of duration that is "pure heterogeneity" (Bergson in Cunningham 1914, 526).

16 Sorrenson (1996, 222) writes: "Although I am arguing that some vehicles can be thought of as scientific instruments, clearly not all can. What, for example of Humboldt's donkey, which transported him and his scientific instruments up mountainsides in South America? Can that vehicle profitably be thought of as a scientific instrument? Clearly not, for the following three reasons. First, Humboldt's donkey had no authority. It did not matter what kind of donkey it was, or whose it was, or how it found its way from one point to another; in contrast, it mattered very much what kind of ship was chosen for a particular voyage, who had commissioned it, and what kind of scientific instruments and techniques made certain its navigation. Second, the donkey left no traces on the maps Humboldt produced; the ship, as I will demonstrate in this paper, most certainly did. Third, the donkey did not offer Humboldt a superior, self-contained, and protected view of the landscapes and civilizations he viewed; the ship, on the other hand, offered all three to its inhabitants." I include this lengthy passage here because the comparison between the donkey and the ship as carriers, and why

they are in fact not similar, raises the question of how a donkey (or other animal) might be a scientific instrument. As an aside, the United States, France, Japan, China, and the USSR can all be said to have used animals (monkeys, dogs, flies) as "scientific instruments" in their exploration of space. Could we then call Laika (the Soviet space dog, 1954–1957) a geographer?

17 The World Ocean Observatory is a "universally accessible, Internet-based place of exchange for information and educational services about the ocean defined as an integrated global social system" (Niell in WOO 2016).

18 I discuss the space of the coastline in geographical thought further in Chapter 6.

19 Similarly, Harvey and Riley (2005, 269) write about the "continued reliance on the expert interpretation of material artefacts" for the interpretation of landscape (see also Driver 1988).

20 "A world [. . .] promised to individuality, the provisional passage and the ephemeral."

21 Similarly, on maps of the United States, the state of Hawai'i is usually inset too, lest the map need to represent most of the Pacific Ocean.

22 List of countries in order of decreasing EEZ areas: United States (11,351,000 km²), France (11,035,000 km²), Australia (8,148,250 km²), Russia (7,566,673 km²), Japan (4,479,358 km²), New Zealand (4,083,744 km²), United Kingdom (3,973,760 km²), Brazil (3,660,955 km²), Canada (2,755,564 km²), India (1,641,514 km²), Argentina (1,159,063 km²), Madagascar (1,225,259 km²), and China (877,019 km²) (OECD 2008, 47).

23 The first accepted limit for territorial waters was three miles, because that was the range of a cannon. A country could claim sovereignty of waters if it could protect them.

24 While the Channel Islands are not British (they are self-governing bailiwicks that are British Crown Dependencies), charts of the Channel trace the boundary as United Kingdom/France.

25 The Minquiers, an uninhabited archipelago, are part of the British Channel Islands located between Jersey and the French Chausey Islands. They have been used as a shelter for fishermen or to cache smuggled goods. The ICJ-CIJ ruling in favor of the United Kingdom followed the 1066 Norman conquest of England's historic link between the Minquiers and Britain, whereas France based its defense on the French taking back Normandy in 1204. However, the treaties (Paris in 1259, Calais in 1259, and Troyes in 1420) that followed did not mention the archipelago.

26 In terms of ocean-space, these colonies yield great advantages with regard to EEZs.

3 1492 and the discoveries of America

Whereas the previous chapter considered the ocean-space in a generalist way, which sought to remain originally neutral, this chapter adopts a more geographically focused approach since it examines, in particular, the early modern European era. Furthermore, this will concentrate this study of the ocean-space to the Atlantic Ocean, since that is the ocean that received most European focus during the early modern era. By studying European charts dating from the early-fifteenth century onward, this chapter will therefore explore how these sought to incorporate the new boundaries of the Atlantic Ocean and the inception of the Pacific Ocean as an independent entity. Alongside maps and surveys, globes will also be discussed since these exemplify the problems of integrating oceans into previous conceptions of the earth. Furthermore, this will bring to the fore a number of issues pertaining to knowledge production and representation. At the core of this period, however, and crucial to the ocean-space itself, this chapter will discuss what is arguably the most important geographical revelation of the Age of Discovery, that is, that all the earth's oceans are interconnected (Parry 1981, xi). This discovery was both physically and intellectually important for the geography of the ocean-space and epitomizes many of the challenges regarding its study. Overall, then, this chapter will consider the changes that occurred in thinking geographically about the earth during the early modern period, focusing in particular on the new places of the ocean-space in geographical thought and how these shaped new visions of the earth on the whole.

As seen in the previous chapter, engagement with the ocean-space was limited until the late-Middle ages, even in coastal nations. The ocean-space was one to be crossed swiftly, not lingered upon. By the late-fifteenth century, however, the warring geopolitical climate, the advent of colonization, and a scientific interest in explorations incentivized interest in the oceans. In particular, alternatives to overland roads were being sought to reach India from Europe, and it is under these circumstances that Christopher Columbus sailed westward from southwestern Spain in 1492 with the intention of finding a shorter route to India. This journey, rather than ending in India, famously landed in the Bahamas, thereby making it the first certainly documented transatlantic journey and acquiring for Columbus fame as the discoverer of America. Furthermore, this voyage, the first of Columbus's four, was an event of global importance within the context of world history, being significant beyond its geographical and political setting. While several claims of precedence exist, Columbus's trip is unique and globally important for several reasons. First, it was his trip that triggered sustained transatlantic activity. Indeed, while there probably had been transatlantic contacts and exchanges prior to 1492, it is Columbus's return to Europe with news of a faraway land that led to major population movements and triggered global transoceanic journeys. Second, it turned an otherwise contained tripartite European/African/Asian, biblically limited worldview into one of limitless unknown. In this sense, upon returning to Europe in 1493 and recounting his travels, despite being incorrect about the geographical nature and location of the places he had encountered, Columbus changed the nature of the known world itself. Finally, Columbus himself remained convinced that he had reached Asia, not a New World, as would slowly emerge. Columbus repeatedly claimed that he had not discovered a New World, but rather a new way to the Old World. This fact, and the difficulty that Columbus and his contemporaries would have in accepting its falsehood, illustrates the challenges linked to changing one's worldview. The notion that Columbus had in fact landed in a place that was then still unknown to Europeans was deduced at a later stage by Amerigo Vespucci, as I will discuss shortly, and it would take more time afterward for this fact to become widely accepted. Together, these points (the transatlantic trigger, the new earthly model, and Columbus's inability to recognize where he had been) reinforce the importance of this particular journey within the context of world history. Thus contextualizing Columbus's journey raises issues regarding the nature of new knowledge, which are crucial to address before continuing further.

Thinking about local events and how they can be conceived of as having global ramifications and philosophical implications points to a number of epistemological issues if we think in terms of the ocean-space and global geographical knowledge, in particular in relation to discovery. It is not here denied that Columbus was indeed the first European man to land in the Americas, but it is certainly problematic to attribute to him an idea that he himself denied. Zerubavel (2003) thus asserts that it is in fact problematic to attribute the discovery of the Americas to Columbus. As far as Columbus never realized that he had reached a new world rather than Asia, as far as Columbus *himself* did not know, or even hypothesize, that he had discovered the New World, Zerubavel argues that Columbus cannot be given credit for a discovery of which he was wholly unaware. He writes:

> My main contention is that "discovering" entails a rather critical *cognitive* dimension, as it implies some *understanding* of what one has "discovered." By giving credit for discovering the New World to someone who stubbornly kept insisting that it was part of the Old, we ignore the obvious fact that [Columbus] did not actually comprehend what he "discovered."
>
> (Zerubavel 2003, x)

This raises the issue of what Zerubavel (2003) terms "the mental discovery of America": How did the Americas become known in European culture and social consciousness, and how was this new land to be dealt with in a scientific and political context? Certainly, "America is both a physical and a mental entity, and the full history of its 'discovery' should therefore be the history of its physical *as well as mental* discovery" (Zerubavel 2003, 35).

While Zerubavel's work is concerned primarily with the mental discovery of the Americas, I will explore how the discovery of America, both physical and mental, is twinned with the discovery of the separateness of the Atlantic Ocean from the Pacific Ocean. Indeed, with the Americas between Europe and Asia, the ocean that was meant to link Europe to Asia was divided into two, with the Americas acting as a continental partition. The mental discovery of the Americas, when it would occur, would then also trigger the discovery of two oceans where only one was thought to lie: the discovery of the Atlantic Ocean and that of the Pacific Ocean as the expanse of water that filled the space between the Americas and Asia. European consciousness and geographical discourse thus had two independent, new, and unknown oceans, which needed to be explored. This is markedly a key moment in the history of the ocean-space, shifting both physical and imagined geographies of the ocean. This is further evidenced by maps and charts of the period, which needed to shift their modes of representations.

Cartographic sources dating from the late-fifteenth century onward visually bear witness to the challenge of making space for two oceans and a landmass where only one ocean had been previously. Specifically, the way in which the Atlantic and Pacific Oceans as well as the so-called New World were mapped from the time of their discovery gives valuable insight as to how these spaces became known both mentally and geographically as they were given space on charts and representations of the world. I will therefore here examine a selection of maps that will bring to the fore these attempts to represent a new conception of the world. In particular, I will consider Martin Waldseemüller's two world maps dating from 1507. These were drawn specifically to accompany a new geographical treatise that sought to update the European worldview in light of the voyages since Columbus and so give particular attention in placing the New World and representing its geographical properties. These two maps are then among the first to purposefully and evidently show the Americas as an entity separate from both Asia and Europe, and to depict the Atlantic and Pacific Oceans in relation to these landmasses. With regard to understanding how knowledge about the ocean-space was produced, these maps are then milestones in that they depict the Atlantic ocean-space and the Pacific ocean-space as such for the first time, representing a New World system and altering a centuries' old model that was thought to be unchangeable. With these maps, the nature of knowledge and knowing is shown to have changed, while the ocean-space gained cartographic space.

Before examining the specifics of the Waldseemüller maps, however, I will discuss different aspects of knowledge from social and cultural standpoints in terms both of epistemology and the history of ideas. This will locate the discussion of knowledge within a framework of history of ideas and history of science, which will be helpful to further put in context these maps and what they represent geographically and epistemologically. While it has been accepted that Columbus triggered a "full-scale cosmographic revolution" by landing in the Bahamas on October 12, 1492, this statement needs to be examined within a larger understanding of knowledge acquisition and the nature of scientific knowledge in order for its implications to be fully grasped (Zerubavel 2003, xii). Furthermore, while it is undeniable from a scientific perspective that the discovery of the Americas signaled progress, where progress is understood as the gradual accumulation of more correct knowledge, it is useful to recognize the implications of treating this discovery as such. Therefore, by questioning ways of understanding knowledge with specific reference to Bachelard, Canguilhem, Popper, and Gigerenzer, Columbus's discovery (or lack thereof) and Zerubavel's contention will be presented in a wider context of knowledge production and cognition. By so doing, we will be able to understand cartographic depictions and, specifically, Waldseemüller's charts of the American continent and the ocean-space in the context beyond the historical setting.

3.1 Knowledge production

Before knowing how to know scientifically, the basis for Western knowledge prior to the Enlightenment was the Bible. Indeed, in Europe at least, prior to the Scientific Revolution, the Roman Catholic Church was a dominating institution over Western civilization as the "pillar and foundation of the truth" (1 Tim. 3:15). This position was not easily challenged, and the power of the Church was such that it both actively and passively halted, or at least impeded, scientific progress. For example, the Church was infamously central in hindering the progress of astronomical knowledge by prosecuting its advocates. In all events, in the early-sixteenth century, any motion toward overturning the authority of the Church and contradicting the truths of the Bible was deemed controversial. This alone makes the Waldseemüller maps uniquely distinctive and important beyond their initial purpose. In this sense, what they represented was more than a particular interpretation of Vespucci's journals: they were in fact overturning the biblical dogma that had dominated Christian thought for fifteen centuries. Thus Waldseemüller's maps are useful here as a point of entry to discuss the nature of knowledge and the manner in which things are known, not judging the facts themselves, but rather examining the nature of the facts. This will help enlighten the nature of knowledge in specific cultural and sociological contexts and as a cultural and sociological construction, as well as how these influence, if at all, geographical knowledge.

The first step toward acknowledging the separateness of the American land-mass was to, perhaps indirectly but fully nonetheless, challenge the widely accepted biblical interpretation about the world and its geography. According to Genesis, following the Great Flood, God divided the world between the three sons of Noah, Shem, Ham, and Japeth: "and of them came the people who scattered the earth" (Gen. 9:19). A widespread and literal interpretation of this passage concluded that the earth had only three continents; this was, in particular, explained in the seventh century by Isidore of Seville's encyclopedic and widespread book *Etymologiae*, which read thus:

> The globe (*orbis*) derives its name from the roundness of the circle, because it resembles a wheel (. . .). Indeed, the Ocean that flows around it on all sides encompasses its furthest reaches in a circle. It is divided onto three parts, one of which is called Asia, the second Europe, and the third Africa.
>
> (Barney et al. 2006, 285)

As Barney et al. (2006, 285) clarify, their translation of "*orbis*" as "globe" emphasizes that "the 'circle' of lands around the Mediterranean" was "the total known extent of land." Nonetheless, this tripartite model was at the core of the European understanding of the world, and this is deeply reflected in the importance of T-and-O maps in the history of cartography. These charts, as well as depicting the physical earth as it is described by Isidore, seek to incorporate theological knowledge. In this sense, they could be used as narrating devices as well as for geographical tools representing the cosmography of the world. Their theological aspect is, furthermore, shown through biblical symbols: Jerusalem is usually prominently in the middle; the Garden of Eden, with Adam and Eve, overlooks the earth; Africa is inhabited with monsters (as extrapolated from Gen. 9:26–27); and the world (flat) is circled by a single ocean. The combination of geographical and religious knowledge on a single chart blurs the two different sorts of knowledge represented, and the interpretation of a certain biblical passage therefore powerfully shapes the understanding and visualization of the earth as a physical entity over several centuries.

When Zerubavel talks of the "cosmographic revolution" caused by the discovery of America, what he is then referring to is a multifaceted epistemic schism that challenged not only the Church, but the nature of knowledge itself. This shock is perhaps comparable to the heliocentric revolution, which was to "re-place the earth" decentering it to an orbital path (Withers 1995, 140). As Porter (in Withers 1995, 140) writes, "the heliocentric revolution (. . .) had dislodged the earth from its central (albeit lonely) place in cosmogony and cosmography (. . .) philosophies which hinged on cosmic influences and analogies were waning." Likewise, if the Americas and the Atlantic and Pacific Oceans were to exist intellectually, this meant that the biblical, tripartite system was not geographically accurate and that for the purposes of geographical and scientific knowledge, a tripartite representation of the world was incorrect. At this point, then, for geographical accuracy to become possible, the dominant system of knowledge had to be revisited and a centuries-old cultural paradigm shifted. Only once this was done could the earth and visual imaginations of the earth resemble one another more accurately.

Scientific progress, in Western method, occurs when "science (or some particular field or theory) (. . .) shows the accumulations of scientific knowledge; an episode in science is progressive when at the end of the episode there is more knowledge than at the beginning" (Bird 2007, 64). Therefore, the discovery of the Americas and the Atlantic and Pacific Oceans constitutes scientific progress:

knowledge about the geographical properties of the earth was increased, and indeed became more accurate. A truer picture both of land-space and the ocean-space was established, even if it clashed with the dogmatic paradigm in place at that point. Nevertheless, the ocean-space, as it is known to be today, and knowledge thereof was acquired and, though it took time to prove that the Americas were indeed separate from Asia, this notion was enough to highlight the fallacies of the old tripartite model and its inherent limitations and inaccuracies.

While Bird (2007, 79) notes that "the history of science is marked by the accumulation of knowledge," the history of humanity is one that can analogously be said to be driven by the need to deal with the implications of new knowledge in relation to past cultural modes. This follows Kuhn (1962, 111) for whom, over the course of the history of science, "when paradigms change, the world itself changes with them." He continues:

> Led by a new paradigm, scientists adopt new instruments and look in new places. Even more important, during revolutions scientists see new and different things when looking with familiar instruments in places they have looked before.
>
> (Kuhn 1962, 111)

This means that new knowledge needs to be both accumulated and understood in an epistemological sense (Bird 2007, 84). Therefore, even if Waldseemüller was articulately displaying in his charts what Vespucci might have lightly thought about, he probably did not, or could not, grasp the full implications of his maps, since they were, literally, changing the world and geographical imaginations about it. His representation of an improbable hypothesis, in fact, was so improbable that it took until the late-eighteenth century for its full extent and implications to be grasped, when the separateness of the Americas from Asia was confirmed when Vitrus Bering sailed the straits that now bear his name. This would not happen until 1728 when the decisive piece of evidence was provided to finally and physically ascertain that the Americas were not an outgrowth of Asia; Vespucci's conjecture and Waldseemüller's representation were thus validated (Black 2003, 71). At the same time, and most importantly with regards to the ocean-space, Bering also honed the shape of the Pacific Ocean and effectively produced more knowledge about the global ocean-space.

The discovery of the Atlantic and Pacific Oceans as two independent (though not separate) oceans as they are known today can therefore be construed a revolutionary scientific hypothesis. This is strongly evidenced when considering the narrative of geographical knowledge production about the ocean-space as a whole and with historical perspective. Certainly, the ability to look back is key here in understanding what happened to the geographical imagination of the ocean-space. This follows Bird who writes:

> Far from being internally accessible, like many of the best things in life, the most exciting contributions to progress are often recognizable as such only with the benefit of hindsight.
>
> (Bird 2007, 87)

Being able to look over the events of two centuries, it becomes possible to understand the production of knowledge about the ocean-space both as making space on a chart but also within a geographical reconstruction of the earth as part of a complex epistemological process. Moreover, these complexities are apparent when considering solely the surface of the ocean-space; specific materialities and spatialities have not been considered fully yet, only taken as part of the notion of the ocean-space. I will address this shortly, though I will first focus on the nature of scientific knowledge, considering the ways in which the history of science is a convoluted succession of sporadic attempts to understand the intellectual background against which epistemological revolutions can take place. This discussion will then allow me to address the Waldseemüller charts from a more intellectually grounded point of view.

Gigerenzer (1991, 254) writes that, "the study of scientific discovery—where do new ideas come from?—has long been denigrated by philosophers as irrelevant to analyzing the growth of scientific knowledge." He then argues that, by his count, there are two principal competing ways of conceptualizing scientific inquiry: Leibniz views it as "an ocean, continuous everywhere and without break or division" whereas Reichenbach divided "this ocean into two great seas, the context of discovery and the context of justification" (Leibniz in Gigerenzer 1991, 254). In all events, in 1507, the idea that the World Ocean was physically discontinued by way of the American continent was introduced into European consciousness. Waldseemüller's maps were certainly, as Barber (2005, 80) writes, "an educated guess and created far ahead of the discovery that the continent stood alone and independent of Asia." Nonetheless, the ideas that they contributed to geographical debates cannot be underestimated, as seen earlier. While Waldseemüller's hypothesis turned out to be correct, the process of its discovery is vague, and this is important in terms of knowledge production about the ocean-space. Indeed, the discovery appears to "possess more structure than thunderbolt guesses but less definite structure than a monolithic logic of discovery" (Gigerenzer 1991, 255). Effectively, Waldseemüller appears to have represented on his map what he extrapolated to be correct from the sources he had available: the journals of Marco Polo and the accounts of Vespucci. When he inferred the existence of the continent, Waldseemüller also exemplified that "discovery is *inspired by* justification": something needed to justify the discrepancies between differing and even contradicting sources (Gigerenzer 1991, 259). The invention of the Americas and the Atlantic and Pacific Oceans responded to a need to make sense of the new information and computing the new data in a new model. In the words of Sherlock Holmes, this translates into his maxim: "How often have I said (. . .) that when you have eliminated the impossible, whatever remains, *however improbable*, must be the truth?" (Conan Doyle 2004, 57). It thus transpires, in Brunswick's terms, that mankind is an "intuitive statistician" (Gigerenzer 1991, 258). Yet whether Waldseemüller statistically deduced the existence of the Americas is of interest only in that it illustrates the problems of inventing new ideas; but as Gigerenzer (1991, 262) concludes, "good ideas are hard to come by, and one should be grateful for those few that one has, whatever their lineage." Nonetheless, while most of the literature is strictly focused on the making of scientific theories, I deem them useful here to simplify the understanding of science to that which can be proved by reasonable, testable methods. Therefore, the term *scientific* is here loosely used, putting emphasis on the importance of verification rather than the Cartesian method behind it.

Bachelard is among the first to attempt to make sense of the parallel story of modern science and philosophy with regard to thought processes. He writes:

> [Scientists] therefore think of the philosophy of science as a review of general results of scientific thought, as a collection of facts. Because science is still unfinished, the philosophy of scientists always remains somehow more or less eclectic, always open, always basic. Although positive results remain, of sorts, poorly coordinated, these results can be delivered as such, as statements of the scientific *state*, to the detriment of the unity that characterizes philosophy. *For scientists, the philosophy of sciences is still concerned with facts.* (. . .)
>
> [On the other hand] the philosopher simply wants science to provide *examples* to prove the activity of spiritual harmony, but believes that s/he has, without science, before science, the power to analyze this harmonious activity. (. . .) It is thus that, all too often, from the pen of the philosopher, relativity degenerates into relativism, the hypothesis into assumption, axioms into primitive truths. In other words, excluding himself from the scientific spirit, the philosopher believes that the philosophy of science can limit itself to the *principles* of science, to the overall themes, or even, by limiting themselves strictly to principles, the philosopher thinks that the purpose of philosophy of science is to link the principles of science to the principles of pure thought, which could turn away from problems of selective application. *For the philosopher, philosophy of science is never entirely the reign of the facts.*
>
> (Bachelard 2005, 2–4)

What this passage highlights is the discrepancy that exists between *doing* science and *thinking about* science, between making new facts and analyzing the making of new facts. Indeed, the philosophy of science is, at times, alien from the science that it studies and vice versa. Bachelard's work is very useful for making sense of the way in which "thinking scientifically, it's placing oneself in the epistemological field halfway between theory and practice" (Bachelard 2005, 5). New ideas and theories that are the result of reasoned and methodological thought processes, therefore, must be situated within the realm of possibility, limited by wider theories and the possibility of practically understanding them. Yet there is a way of making sense of the improbable, or at least that which is not explained by previous models, that Bachelard discusses. He uses the example of the concept of negative mass, which was made sense of by use of dialectic philosophy, or what he calls the philosophy of the "Why not?" (Bachelard 2005, 36). He writes:

> Why wouldn't a mass be negative? In which framework of experiments could we discover a *negative mass*? What is the characteristic that, in its propagation, results in a negative mass? In short, the theory holds well, and, for the

price of a few basic modifications, it doesn't hesitate to seek the realization of a brand new concept, with no roots in common reality.

(Bachelard 2005, 36)

In effect, the existence of negative mass is impossible within the framework of Newtonian physics, but once the limits of the framework are altered, negative mass can be turned into a viable concept. The mental discovery of the negative mass is, as in the case of the Atlantic Ocean and the American continental land-mass, just as challenging as the invention of the concept itself: the way to know things needed to be reinvented. As such, Bachelard (2005) goes on to explain what he calls "the philosophy of the no." He writes:

> Generalization by the no must include what it denies. In fact, the rise of sci-
> entific thought over the past century comes from such dialectical general-
> izations that encompass what they deny. Therefore, non-Euclidian geometry
> encompasses Euclidian geometry; non-Newtonian mechanics encompass
> Newtonian mechanics; wave mechanics encompass relativistic mechanics.
>
> (Bachelard 2005, 137)

Essentially, Bachelard's own theory of the history of science is concerned with making sense of past understandings of the history of science itself, and proposes a model by which understanding science also encompasses the lack of under-standing of science. As he concludes,

> In effect, what we take away from the image has to be found again in the
> rectified concept. We therefore gladly say that the atom is exactly the *sum of
> the critiques* to which its first image was subjected. Coherent knowledge is
> the product not of architectural reason but of polemic/controversial reason.
>
> (Bachelard 2005, 139)[1]

What Bachelard argues in favor of, therefore, is an understanding of science that, while acknowledging specific shifts in thought processes and the invention of new ideas, recognizes that every precept has an opposing one: every idea has an oppo-site, which may or may not one day be enunciated and that may or may not one day become the norm. This approach to science and knowledge is very useful with regard to making sense of the existence of the Atlantic and Pacific Oceans and the Americas because it suggests that the idea of a non-tripartite world was part of the idea of a tripartite world. While this does not mean that the idea was easier to conceive of or accept, it may help to make sense of it in a philosophical and epis-temological way, as will be discussed further in a moment. First, though, I will examine other theories of the history of science from a philosophical standpoint, to broaden our understanding of the knowledge production about the ocean-space.

Canguilhem sought to define an approach to the philosophy of science that was both rigorous and practical. While he principally worked with examples from the field of biological sciences, his ideas are transferable to the general sciences. He frames the purpose of the history of science as seeking to understand scientific thought as more than just a succession of scientific moments, but instead holding a larger narrative. He writes:

> Should we write the history of science like a special chapter of the general history of civilization? Or should we look for, in the scientific conceptualizations of a given moment, for the expression on the general state of mind of an epoch, a *Weltanschauung*? (. . .) Is this history then within the realm of the historian as exegete, philologist and erudite (especially with regard to Antiquity) or of the scholar who is able to dominate the problem of which he recounts the story? Must he be capable of making the scientific question progress so that he may be able to understand its past through the clumsy attempts that helped formulate it? Or is it enough to manage to highlight the historical, almost outdated, aspect of old ideas and concepts to reveal the out-dated character of the ideas despite the permanence of the terms used. Finally, and following what precedes, what is the value of the history of science to science? Is the history of science no more than a museum of mistakes of human reason, if truth, the result of research, is left out? In this case, the history of science would not be worth an hour's troubles because, from this point of view, the history of sciences is the history of that which is not scientific.
>
> (Canguilhem 1980, 43)

The history of science, Canguilhem maintains, traces the human understanding of the world through the mistakes of its reasoning. It is a succession of righting wrongs and adjusting new discoveries to old systems. Truth, or what is understood as truth in effect emerges only as a consequential product of various attempts to make sense of the world. As such, he continues:

> The advantage of a well-understood history of science seems to reveal history in science. History, that is according to us, the sense of possibility. Knowing is less coming up against a reality than validating a possibility by rendering it necessary. (. . .) The illusion could have been a truth. The truth will reveal itself one day maybe an illusion.
>
> (Canguilhem 1980, 47)

What constitutes truth and what seems to be an illusion is, therefore, less a question of actual scientific knowledge and more concerned with the cultural basis for knowledge. When new things are discovered, beyond the scientific implications that they have, Canguilhem puts emphasis on the place of cultural and social beliefs. He summarizes this as follows:

> We will say that life must be the succession of unpredictable conversions, forcing one to seize a possibility whose meaning is never so clearly revealed for our understanding than when it is disconcerting.
>
> (Canguilhem 1980, 39)

Whether Canguilhem speaks of the life of an individual or that of a society is not important here, as it can be argued that the one mirrors the other, and vice versa. Regardless, this idea that the nature of knowledge is closely relatable to the general beliefs of society is fundamental here within the context of the mental discovery of America. The invention of America and the split ocean-space that it entails can thus be equated to an extraordinary ability to process new ideas, even if they bring forth previously unimaginable new truths.

The final notion on the nature of knowing to be considered here will be that of Popper. Popper is among the most eminent and prolific recent writers on the philosophy and history of science. His position on the need to study the history of science is very clear: he propounds that "*all the growth of knowledge can be studied best by studying the growth of scientific knowledge*" (Popper 1959, 15, emphasis in original). Popper here equates knowing with knowing scientifically, therefore suggesting that the history of science is the history of things that can be proved mathematically and known critically. This distances his stance from those of Bachelard and Canguilhem, who focused more on ideas in general, only applying their discourse to scientific method. As Popper (1959, 15, emphasis in original) writes:

> there is at least one philosophical problem in which all thinking men are interested. It is the problem of cosmology: *the problem of understanding the world—including ourselves, and our knowledge, as part of the world.* All science is cosmology.

Here Popper articulates his earlier point further, stating that the question of understanding science historically is equivalent to knowing the world itself. Something is not known until it is scientifically proven. He argues further:

> from a biological or evolutionary point of view, science, or progress in science, may be regarded as a means used by the human species to adapt itself to the environment: to invade new environmental niches, and even to invent new environmental niches.
>
> (Popper 1994, 2)

Popper thus conceives of theories about the world as ways of dealing with the world and scientific revolutions as ways of dealing with a changing world. A new theory is a new way of dealing with novel and unfamiliar information that does not fit within an old model. Scientific (knowledge) revolutions are thus perceived as ways of incorporating new information.

In order to have a scientific revolution, one therefore has to be able to see beyond one's own beliefs in order to obliterate them and replace them by a dogma that is more correct, that tends more toward truth, or at least a new understanding of truth. The ability to manufacture a new way of understanding the way things appear to be (which is beyond the remit of the previous model) is what is revolutionary about a scientific revolution. Consequently, as Bernard (in Popper 1994, 7) notes, "those who have an excessive faith in their ideas are not fitted well to make discoveries." As such, a new theory about the Atlantic Ocean and the Americas emerges as being a human way of adapting Europe to new information, and this was possible because Vespucci and Waldseemüller were able to think beyond the limitations of the times in which they lived. The cosmographic shock did not impose itself upon a passive European consciousness; rather, it was concordant with certain events in Europe, such as the publication of the 1507 maps, which, in turn, were part of a massive overhaul of knowledge and reorganization of cosmography in a shocking way. Following Popper, then, Zerubavel's shock transpires as being more of a process whose conclusions are shocking rather than a shock unto itself.

It must be noted, however, that Popper makes a fundamental differentiation between scientific and ideological revolutions. The knowledge revolution that was set in motion by the discovery of the Americas, physical and mental, was, of course, of both sorts: as well as constituting a huge knowledge leap, it challenged existing ideologies. Along with the Copernican and Darwinian revolutions, the "American, Atlantic and Pacific" one was *"ideological* in so far as [it] changed man's (*sic*) view of his place in the universe" (Popper 1994, 17). Nevertheless, it was scientific in that it overthrew the tripartite (geographically inaccurate though biblically faithful) model of understanding the world.

Examining how these several ways of conceiving the notion of changing knowledges and questioning what it means to know things in relation to how a predominant dogma might change is important here to put in context the discovery of the oceans-space as it is known to exist today. These insights into the philosophy of science and knowledge are therefore helpful with regard to making sense of the mental discovery of the Atlantic Ocean and the Americas. In particular, these notions have consequences in relation to how knowledge can be made at a distance and travel, and the philosophy of science and knowledge production brings to the fore the tight entwinement between cultural settings and practical applications. Thus the connection between social links and scientific knowledge is emphasized. Further, by considering the question of the cognitive mechanisms behind the making of new ideas, the mental discovery of the Atlantic Ocean (and the Americas) emerges as a geographical event, which is representative of an epistemological paradigm shift.

3.2   The existence of a New World

Although many features of the earth's physical properties remain unknown, or even questionable, certain facts are today undeniable. Certainly, it is nonsensical to refute the existence of the Americas or the geographical fact of the Atlantic and Pacific Oceans. These realities, however, had to be discovered and understood before they could be factual. The voyages of the Age of Discovery were instrumental in these processes, since, as Parry (1981, ix) analyzes, it was a period that

> revealed features of the globe formerly unknown, not only to Europeans, but to mankind (*sic*) as a whole. Of these, the most important was the unity of the

sea: the geographical fact that the salt seas of the world are connected; that all the countries possessing sea coasts are mutually accessible by sea.

Columbus's journey was an initial step toward discovering this fact and its complexities: he had correctly assumed that the ocean would take him to India, but failed to realize just how difficult and indirect this would be. When Columbus returned from the Indies—west in reality, east in his mind—Europe had yet to grasp the geographical reality of the New World, as discussed earlier, but also re-draw the ocean-space in a more elaborate way. As Parry (1981, xi) writes:

> Knowledge of the continuous sea passages from ocean to ocean round the world—established knowledge, not geographical hypothesis—was the outcome of the century or so of European maritime questing (. . .). Until the last quarter of the fifteenth century most of these passages were unknown to Europeans, and some were absolutely unknown—unknown to anyone. The only major connecting passages known and regularly used by European shipping were the straits which join the Mediterranean and the Baltic with the North Atlantic, and the Black Sea with the Mediterranean. No ship had penetrated the Caribbean (. . .). On the opposite side of the world, Chinese, Malay and probably Arab navigators were familiar with the long tortuous passages, threading between the islands, which connected the Bay of Bengal with the China Sea, and so, more remotely, the Indian Ocean with the Pacific; but none of them, so far as we know, had ventured far into the Pacific. That was a maritime world of its own: to Europeans, its very existence as a separate ocean was unknown, and so was the great continental barrier of the Americas which divides it from the Atlantic.

It is difficult to understate the importance of the mental invention of the Americas and the division of the World Ocean. As a result of European exploration, the geography of the earth was redrawn and paradigms of knowledge production shifted. While the physical discoveries were momentous, the mental leap was phenomenal. Within one hundred years of Columbus's return, geographical knowledge of the earth had been revolutionized, not only in practice but also in theory: the earth's characteristics were now to be discovered by exploration and confirmed according to a new paradigm. However, while the tenets of knowledge changed, the practicalities of science and geographical science across the earth's surface remained technically and practically challenging. I return to discussing this with practical examples relating to the Atlantic and Pacific ocean-spaces in the following chapter; for the moment, I focus on the generalities of knowledge production about the ocean-space with examples pertaining to the Atlantic and Pacific Oceans.

Making sense of the mental discovery of the Atlantic Ocean and its Pacific counterpart is central to understanding the ocean-space, because the ways in which new knowledge and new forms of knowledge were recognized and then represented provides useful insight into the sociocultural contexts of the time and has implications for how we can understand knowledge production. For the cartographers who sought to map the new and emerging world, or even just new and emerging conceptualizations of the world, understanding and representing both the past and the present was a complex task. New maps still had a connection to previous ones, and, following Bachelard, cartographers were bound to represent both. Maps thus emerge as barometers to measure the changing world and the reworked notions about the world and the physicality of the earth. In fact, Thrower (1981, 1) argues that "as a branch of human endeavor, cartography has a long and interesting history that well reflects the state of cultural activity, as well as the perception of the world, in different periods." Bassin (1999, 109–111) furthers:

> [Maps] are artifacts of the political and social mentalities of the civilizations and epochs that produced them. (. . .) The cultural significance of maps—in this case exclusively world maps—is also emphasized by Whitfield, who likens cartographic representation to art and literature in that all come from the same impulse to crystallize, comprehend and control aspects of reality.

The importance of the cartographic representation, and the parallel evolution of the American concept and the ocean-space is therefore crucial to the understanding of the "mental discovery" of the new world geography or even New World geography. Certainly, Zerubavel (2003, 104) notes:

> Many of the maps made during the sixteenth century thus embody the various efforts made by Europe to reconcile the extremes of innovation and denial in its overall response to the understandably traumatic discovery of America.

The representation of the New World and the representation of the world in its new shape therefore emerges as key in understanding how the Americas and the Atlantic and Pacific Oceans came to be known. The mental process of discovery of these as new places and not new parts of old ones followed the first actual European encounter with these places, yet it is the intellectual event that reorganized the earth's geography. "[As] everyone knows, on October 12, 1492, Columbus made a historic landfall on the tiny island of Guanahaní in the Bahamas. America, however had yet to be discovered" (Zerubavel 2003, 118). The ocean-space, too, was discovered, now being known according to new systems and within a new worldview. Both would now need to be represented.

3.3 Representing the New World Order

While the exact nature of the recent Spanish and Portuguese faraway discoveries was still being debated in European circles, as it would be for decades to come, a small group of German clerics and humanists from Saint Dié sought to make sense of the kinds of lands and places that travelers were describing and speculated on how they might geographically fit in the world. Their primary source of knowledge about the great journeys were the widely published journals of Vespucci in which he recounted his 1501–02 trip toward the Americas (Ehrenberg 2006, 68). After traveling down the eastern coast of South America, Vespucci (in Ehrenberg 2006, 68) wrote in his journal that he thought that he had arrived to a "new land," which he "observed to be a continent."

The group of Saint Dié, under the leadership of a Matthias Ringman, sought to integrate Vespucci's account and produced, afforded by the patronage of Duke René II of Lorraine, two maps that challenged assumptions made about the geography of the world throughout Western thought. The maps were drawn by Martin Waldseemüller and published in 1507 accompanying a treatise called *Cosmographiae Introductio*, which was an attempt to modernize geography and update the classical works of Ptolemy. Of the two maps, one was projected in a pseudo-conical projection, the other was drawn like a flattened globe, giving more visible importance to the size of the earth and the relative sizes of lands. The former was called *Universalis Cosmographia secundum Ptholmaei Trasitionam et Americi Vespucii adioruque lustrationes* (*A Drawing of the Whole Earth According to the Tradition of Ptolemy and the Voyages of Amerigo Vespucci and Others*) (Figure 3.1), and is known as Waldseemüller's Conical Map. It measures 1.20 meters by 2.25 and was printed from twelve woodblocks. The latter, smaller, is referred to as Waldseemüller's Globular Map of the World and depicts the new relative sizes of the earth and its (now) four continents (Figure 3.2). Only four copies of it remain in existence. Both maps use the name "America" on the southernmost landmass, being the first instances of the use of the name to designate the new continent.

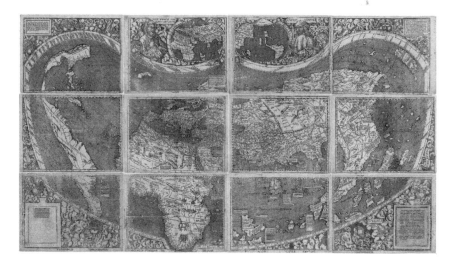

Figure 3.1 Waldseemüller's Conical Map of the World, drawn in 1507, according to the tradition of Ptolemy and the voyages of Amerigo Vespucci and others.

Source: Library of Congress, Geography and Maps division.

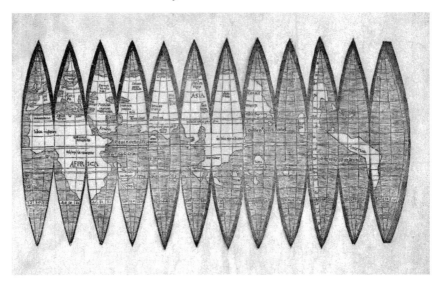

Figure 3.2 Waldseemüller's "Globe Gores" (Globular Map of the World), drawn in 1507.
Source: (James Ford Bell Library), University of Minnesota, Minneapolis.

These maps are the first to represent the American landmass as an entity separate from both Europe and Asia, essentially distinct from the known world and beyond the realm of current geographical knowledge, flanked by two oceanic bodies. Also, by giving the name "America" to the landmass, the maps honor Vespucci, whom Waldseemüller understood to be the discoverer of the continent. This naming of the Americas is relevant only inasmuch as it gave an independent identity to the new continent, not because it praised a particular Italian sailor.

There is, however, a larger issue related to the naming of the ocean-space in these maps, and it has to do with how it differentiates land and ocean-space. Until the rediscovery of Ptolemy's *Geography* by Europeans in 1409, cartographic practices strictly differentiated between land and the ocean-space. This is visible on mappaemundi on one hand, and portolans on the other. The former were concerned with journeys on land, whereas the latter were practical representations of oceanic paths. As Steinberg writes, "mappaemundi and portolan charts (. . .) each came to produce elemental distinctions between land and sea" (Steinberg 2009, 478). Nevertheless, the Waldseemüller charts begin to blur the boundaries between land and sea by representing them and naming them equally. Indeed, the difference between the two surfaces are lessened, following Ptolemy, and the charts instead focus on representing space itself. Therefore, the binary opposition between land and ocean is temporarily erased as these maps focus instead on the depiction of space as an absolute, devoid of cultural assignations, which hierarchize kinds of space. By so doing, Waldseemüller succeeds in bringing to the fore the importance of a global picture of relative geography where inserting a new continent has consequences for oceanic divisions. As will be demonstrated in the following chapters, however, this representational equality is short-lived and limited to this period's charts, as the ocean-space later becomes construed as a zone of transit to be crossed. This dichotomy is then further exacerbated with the advent of deep-see exploration, as will be discussed in Chapter 5.

Returning to Saint Dié, however, the charts produced there are very significant and historically important: these were not the only attempts to incorporate the new knowledges into the old worldview and represent the world as new, as had been the case with previous maps. One such example is Juan de la Cosa's 1500 *World Map* (Figure 3.3), which depicts lands discovered by Columbus, Vespucci, and de la Cosa himself. (De la Cosa had traveled on Columbus's first two journeys to the West Indies.) However, de la Cosa's portolan does not distinctly make the Americas a "new" continent and the Atlantic a "new" ocean, and does not make bold statements about the earth's new layout. Unlike the Waldseemüller map, de la Cosa's focuses on the smaller events of discovery rather than a wider perspective and places emphasis on the nationality of the discoveries. To wit, the discoveries made by Jean Cabot are marked as "land discovered by the English"; for Spain, he writes, "this cape was discovered in 1499 by Vincens Ians for the Crown of Castille" (Barber 2005, 78). In this sense, de la Cosa's map follows a different tradition of cartography, which revolved around nationality, sovereignty, and governance. De la Cosa does not seek to depict anything more than a vision of the world, albeit integrating recent discoveries, which places emphasis on the separation of lands.

Furthermore, by accurately depicting Cuba as an island, not a promontory out of a larger landmass (which is what Columbus himself believed to be the case), this map further consolidates the idea that the map was less concerned with making sense of the greater picture of the geographical reality of the New World and more as a conceited effort to understand precisely what Columbus, Vespucci, de la Cosa, and Cabot saw with regard to what was already known, that is Europe, Asia, and Africa. De la Cosa's map was not revolutionary ideologically or epistemologically; rather, it solely adapted the old model to new knowledge. Furthermore, unlike the Waldseemüller maps, de la Cosa's map is drawn with practical, navigational purposes in mind, as demonstrated by the presence of rhumb lines. These reemphasize the map's link with the past and both its scale and its impact find themselves reduced as a consequence of this.

Having examined these three seminal charts concurrently, the differences between de la Cosa's work and Waldseemüller's charts are made clear, and these reflect divergent approaches to making sense of new geographical knowledge. Through shifting the borders of the known beyond anything that was previously imaginable, the Globular and Conical maps assimilate new knowledge and represent a shifted paradigm of knowledge creation, one that dramatically changed the shape of the world. Likewise, when Vespucci had thought that the coast along which he had traveled southward was not attached to any known mass, he concluded that he must have navigated down a new continent. He too was reshaping the earth and our understanding of it. At the same time, it is also noteworthy that it is in itself phenomenal that both Vespucci's and Waldseemüller's guesses

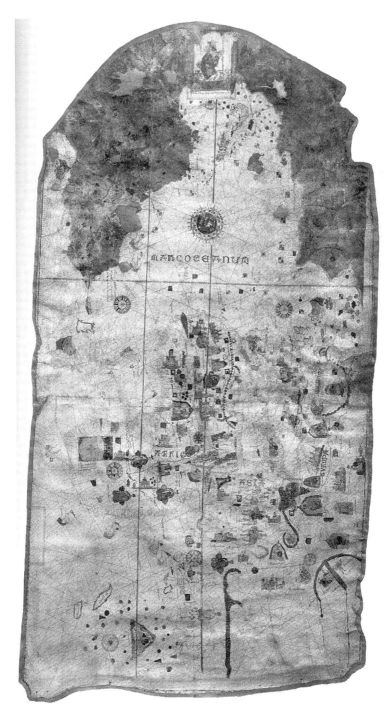

Figure 3.3 De la Cosa's 1500 *World Map*. The original parchment of this map was printed on a piece of ox hide measuring 96 × 183 cm.

Source: http:// en.wikipedia.org/wiki/Map_of_Juan_De_La_Cosa

about the geographical independence of wherever it was Columbus had landed on proved right.

As previously noted, the physical separateness of the Americas and Asia was only confirmed in the mid-eighteenth century by Bering. This was followed by the confirmation by James Cook of Alaska's Americanness in the course of his third journey between 1776 and 1779 during which he mapped the greater part of the west coast of North America (Williams 2002). Until these two events, two and a half centuries after the charts discussed earlier, the possibility of the tripartite model being accurate remained alive. It could only be rejected fully in 1779. It follows that Zerubavel (2003, 66) argues that this is the year of the full discovery of America; that is, its physical as well as mental discovery:

> Only when Alaska's American identity was thus firmly established by Cook could the possibility that Asia and America might somehow be connected be ruled out once and for all. Only then, 286 years after Columbus' landfall in the Bahamas, was the mental discovery of America by Europe fully completed!

While the physical geographical reality was ascertained, however, the sociocultural one was more problematic. While the actual separateness was the "theme underlying the long history of the mental discovery of America," what is of interest here is what this separateness means for the representation of the ocean-space on maps and in European consciousness as something different to what it had been assumed to be (Zerubavel 2003, 50).

The stretch of water that extended westward beyond the Strait of Gibraltar had been assumed, since Antiquity, to be a river that circled the earth. When the earth was known to be spherical, and there were no hints in Europe as to the existence of the Americas, this same expanse of water logically extended into the pathway to Asia. The circular, limiting ocean of Antiquity became a global expanse that linked two distinct parts of the world. The size of this body of water, though, was unclear. Columbus purportedly adapted figures so that they would support his demand for funds from Isabella and Ferdinand, thus making the distance from the Canary Islands to Japan a mere 2,400 nautical miles wide. This was the smallest figure anybody had put forward, the Atlantic Ocean being but a slither between the two. This nearness of Asia and thinness of the Atlantic Ocean was supported, though incorrectly, by Martin Behaim's 1492 Erdapfel that placed Japan 1,500 nautical miles further to the east than Marco Polo had suggested it to be. Effectively, this placed Asia within island-hopping distance of Europe. Under these spatially limited circumstances, it was further unlikely that there might be a continent and a second ocean between Europe and Asia. Therefore, a further challenge faced by the Saint Dié scientific contingent was also to make physical space for the New World on the map, to allow room for two oceans and one continent on what was otherwise a compact and full map of the world. Through doing so, the global ocean-space, in its most basic physicality, was beginning to resemble what it is known to be today. This is what the Waldseemüller maps, unlike de la Cosa's for instance, started doing: shaping the ocean-space on a global scale.

Both of Waldseemüller's maps deal with the question of the physical space necessary for a new continent. The shape and size of the Atlantic and Pacific Oceans and the Americas were drawn into European consciousness. They would only be geographically adjusted and scientifically corrected throughout the eighteenth century, as expeditions would be sent out to specifically measure relative positions of coasts and islands, thereby creating maps of the world whose geography was relatively accurate (Berthon and Robinson 1991). In the meantime, the Globular Map shows that India is roughly equidistant from Asia as it is from the Americas. Japan is estimated to be halfway between the Korean Peninsula and North America. For want of any information concerning the west coast of the new continent, it is represented by a schematic line. Further, the continent has no significant width: it appears as no more than a slither of land. In effect, this first map makes the Pacific Ocean the same width as the Atlantic Ocean, and the Americas are, approximately, half the size of the Indian subcontinent. Nevertheless, the global ocean-space is for the first time shown in a way that is truest about its actual shape.

The Conical Map is, on the other hand, less geographically precise and presumptuous. While it draws Europe, Asia, and Africa (to the best contemporary knowledge), together with the Caribbean, in great detail, the geographical wholeness of the earth is less visually clear. The size of the Pacific Ocean is not as evident to the observer, and it appears haphazardly speckled with anonymous islands. Nonetheless, the most important feature of this map is the very clear and precise eastern coast of the American continent. This outline of the coast was likely compiled from various accounts, including Vespucci's journals and letters and other available sources. If the eastern coast is precise, the western one is entirely unknown and is schematized, as on the Globular Map. Similarly, the width of Vespucci's "new continent" is unknown.

Despite gaps in knowledge, these maps are stepping stones in the direction of mentally discovering the Atlantic and Pacific Oceans and the Americas. In giving the Atlantic and Pacific Oceans a place, both relative and absolute, on these new maps, these oceans become crucial elements in the process of knowledge production about the ocean-space. Indeed, their spatiality and materialities are evidenced in their simplest forms, consolidating the notion of ocean-space. It is thus that undeniable geographical features of the earth were first enacted by the pen (or carving knives) of Waldseemüller and then integrated shifting understandings of the world and changing models of knowledge. In a sense, that the maps also ended up being somewhat accurate is almost irrelevant: the fundamental point is the matter of shifting knowledge paradigms and how new geography of the earth became mainstream.

By depicting the earth as it never had been conceived of before, the Waldseemüller maps are exemplary of both a new knowledge paradigm and an innovative point of view that shows a new global totality and creates a new ocean-space. The maps therefore fit within the "cultural and historical contexts of Western and global images and imaginings" (Cosgrove 1994, 270). As representations of the whole earth from a novel perspective, they are comparable to the Apollo space photographs that Cosgrove (1994) studied. Of these, he writes, "they have been enormously significant (. . .) in altering the shape of contemporary *geographical imagination*" (Cosgrove 1994, 271). By showing the earth from a different perspective, the Waldseemüller maps and the Apollo photographs represent a different way of thinking about the earth and its geography, which is located within wider cultural discourses and, crucially, geographical imaginations. Certainly, Cosgrove (1994, 290) writes that "such a perspective, in addition to immediately localizing us within the world rather than beyond it, might, in John Kennedy's words and Michael Collins's hopes, 'return us safely to earth.'"

I have discussed here in some detail the processes involved in creating and then accepting the new concepts of the Atlantic and Pacific Oceans and the Americas. Their appearance in the Saint Dié work was perhaps the most significant step toward doing this. Even if some publications still denied the separateness of the Americas and carried on advocating a tripartite worldview, these (vain) efforts were limited and did not seriously affect the mental acceptance of the Americas or indeed the undeniable reality of the ocean-space. What was, however, of more concern and held more implications as regards the social and political balance of Europe was the way in which these new lands should be dealt with politically and religiously. If the Bible was to become an unsatisfactory source of geographical knowledge, the manner in which the Church dealt with the new discoveries is useful to understand the sociocultural context of the discovery of the Atlantic Ocean. Therefore, it is useful to reconsider here the role of the 1494 Treaty of Tordesillas.

The principal interests of the Church toward the New World regarded religious supremacy. In 1494, the Treaty of Tordesillas exclusively divided the newly discovered lands, wherever they might actually be, between Spain and Portugal. The assumptions and implications of this were manifold. First, as it was signed in 1494, only one year after Columbus had returned from his first journey, the nature of what it was dividing was very unclear. It is conceivable that the treaty's authors thought that they were parceling off yet unknown parts of Asia rather than a new continent. Either way, the treaty implicitly covered both possibilities by not specifying the nature of the lands it was dividing. Geographically, its only particulars were that all lands to the east of the line of longitude running 370 leagues west of the Cape Verde Islands would be Portuguese. All those to the west would be Spanish. This line fell partway across what is today known as Brazil, thereby making this portion of land Portuguese and the rest of yet-undiscovered South America would be Spanish. This made the Spanish loot potentially immense, depending on the size of the earth. However, the length of the leagues used was not agreed upon, and the exact location of the Cape Verde Islands was unknown, so the treaty's delineations were really no more than a vague approximation dependent upon unfixed geographical locales. In spite of this, the Tordesillas line appears on de la Cosa's 1500 map, in a location that could well be 370 leagues west of the Cape Verde Islands. On this map, the line represents only one interpretation of the treaty's interpretations and the geography of the ocean-space. Several more suggested locations for the line are displayed on the map that follows, which was assembled using several fifteenth century maps (Figure 3.4).

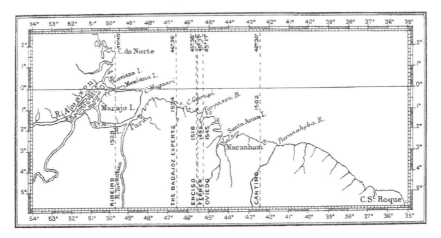

Figure 3.4 Map of the line of demarcation according to the Treaty of Tordesillas. This map shows different interpretations of where the line should be.

Source: https://commons.wikimedia.org/wiki/File:Early_Tordesillas_lines.jpg

By 1529, Europe was beginning to assemble a more complete picture of what was west of the Tordesillas line. Specifically, the idea of a new continent was starting to be established (this was already twenty-two years after Waldseemüller's maps and twenty-four or twenty-five years after the publication of Vespucci's diaries), and Spain and Portugal sought to clarify the geographical implications of the line and its associated claims. Furthermore, the size of the earth was still the subject of much contention. Certainly, the Waldseemüller maps and Behaim's Erdapfel are evidence today that the size of the earth had been vastly underestimated. In 1529, however, the Treaty of Saragossa dictated that the Tordesillas line should continue all around the earth. Together, Spain and Portugal thus appropriated for themselves the entirety of the world: the scope of missionary zeal, in theory, therefore covered the whole earth. Crucially, this was despite the fact that the exact nature of what was being divided remained uncertain; yet significantly, interests were more land-oriented than oceanic. The ocean-space is not yet of interest and is thus not discussed. What is certain, however, is that there was no way of knowing the nature of what was being given to either party. Perhaps unsurprisingly, this brings forth several problems. First, the calculation of longitude was still very precarious, as no viable solution to measure it had yet been established or invented. This is important because being able to measure arcs of the earth would have simplified the task of dividing it according to a line that depended upon longitudinal measurement. Second, some of the far-off lands whose ownership was being disputed had already been claimed by one of the parties involved. This was the case of the Maluku Islands that had been reached by the Portuguese in 1512 but which were, eventually, found to be in the Spanish half of the earth. It is significant that even as the geography of the earth was being made sense of and the concept of a large, four-continent world was being introduced, mentally discovering it is best done if accompanied by ownership. Indeed, after mentally discovering the Atlantic Ocean and the Americas, European powers immediately sought to establish ownership over resources and peoples through religious fervor and imperial enterprise. As noted earlier, the ocean-space thus emerged as an intermediate between the imperial core and its periphery where it was "crucial to the discursive and material workings to the era's European empires" (Steinberg 2009, 481). Certainly, as Adam Smith (in Elliot 1990, 43) noted, "the establishment of the European colonies in America and the West Indies rose from no necessity," being rather motivated by the accumulation of wealth. (See Pohl 1990 for more on the economic aspect of the discovery of the New World.)

What needs to be impressed here, therefore, is that, at the beginning of the sixteenth century, the Atlantic ocean-space emerged in European geographical thought as both a physical space and a discursive idea. Likewise, the Pacific ocean-space, though not yet engaged with, had claimed a physical space on charts and in wider conversations about the size of the earth as a whole. This dual discovery, that is, a physical and a mental one, is redolent of a larger cognitive process that ended up redefining the production of knowledge more generally. Indeed, the ocean-space, here, became known both factually and culturally, and neither could possibly be divorced from the other.

3.4 Conclusion

The manner in which facts become known and then culturally accepted as new models of knowledge is always convoluted. This is especially so when the new model controverts centuries of previously unquestioned and unquestionable truths. With the discovery of the Americas, fifteenth-century Europe found itself having to deal with the debunking of the hitherto totally accepted, biblical, tripartite worldly model. This also meant readjusting the model of the world's oceans and realizing that there were, in fact, two where only one had previously existed, as well as having to revisit calculations about the size of the earth. In fact, the scope of the knowledge revolution that Columbus triggered upon returning from the Americas was unprecedented. Though he cannot be credited with changing the nature of knowledge, in a short period following his return, the nature of knowledge itself changed: evidence and sound conjectures became central. Whereas the Bible was previously taken as true in essence, when Vespucci, Waldseemüller and others began hypothesizing about the separateness of the Americas, the need to be able to prove ideas became paramount. Consequentially, being able to show the individuality of the New World came with the need to change the way that things were known. In turn, the manner in which knowledge and evidence could be made to travel emerged as crucial to this enterprise. These two aspects of making geographical knowledge can be construed as being the first step toward the Scientific Revolution, which took this yet further. In a way, Waldseemüller and the Saint Dié savants seem to have anticipated the new way of knowing, that is, knowing scientifically, by tentatively drawing a new continent.

Alongside this, what is further important to highlight here is the multifaceted aspect of the discovery of the Americas and the Atlantic and Pacific Oceans. At the beginning of this chapter, several ways of knowing and discovering new things were discussed. This was to try and understand the various levels at which knowledge has to be made. In the case of the making of the ocean-space, the convoluted process by which it was mentally discovered is seen on the maps of the period. In the end, the mental discovery of these new oceanic spaces and their representation in the Waldseemüller maps exemplify the fundamental question of attempting to make sense of the earth according to a new system of knowledge where "*seeing* [is] the foundation of knowledge" (Cosgrove 1994, 272).

Note

1 All translations of French-language texts are the author's own.

4 The Enlightenment and the ocean-space

After the so-called European Age of Discovery revealed far-off lands and divided oceans, and as Europeans gradually introduced these into their worldviews and charts, the Enlightenment was the period that sought to comprehend these discoveries from scientific and philosophical standpoints. While the Age of Discovery collected new information about the world, this accumulation of knowledge was mainly disorganized, following the haphazard routes that ships were taking at the mercy of ill-understood winds and currents. Yet as phenomena were increasingly understood and cartographic methods refined, increased knowledge of the world as a whole was allowing ships to sail the ocean with a reasonable idea of where they were going and how they might get there. The Enlightenment therefore saw sailors, travelers, and explorers indefatigably collect and interpret information, which was then used to produce knowledge. In this chapter, the Enlightenment will be considered in relation to the making of geographical knowledge and the history of geographical thought and practice, focusing in particular on how the period's scientists sought to understand and make knowledge about the ocean-space. Building on Livingstone and Withers (1999c, vii), this chapter will examine

> the geography *of* (the) Enlightenment, and the ways in which geographical knowledge was practiced *in* the Enlightenment, but also (. . .) how questions of geographical nature fitted into and were actively engaged with those of matters of rationality and human nature routinely held to characterize 'the' Enlightenment.

While first considering the limitations of periodization, I will define the precise historical period that will be examined and its connection to questions of geographical enlightenment. Crucially, this will adjust the limits of the Enlightenment for the purpose of an ocean-centered discussion, addressing the fact that borders that are delineated according to political events are of little relevance to the enterprise of knowing the ocean-space. Moreover, there is much debate about national enlightenments, such as the French or Scottish Enlightenment, and defining an oceanic or ocean-space Enlightenment will avoid nationalistic debates, which are of little use to the current discussion. This follows Withers (2007, 42) for whom "dismissing *the* Enlightenment in national context as *the* way to understand the Enlightenment geographically is consistent both with that spirit of critical revision now apparent in Enlightenment studies and with that disquiet expressed at the supposedly universal nature of science." In the latter part of this chapter, I will consider specific cartographic projects that highlight certain aspects of ocean-space geography during the Enlightenment. In particular, examining mapping events located in the Atlantic Ocean will bring to the fore the knowledge processes behind the scientific knowledge of the ocean-space. This will contribute to historical geography debates by locating knowledge production within oceanic, scientific, and Enlightenment settings.

4.1 The (Atlantic) Enlightenment

By the mid-sixteenth century, European explorers and navigators had collected a wide range of information about the ocean based on a variety of experiences. These included sailed as well as overland expeditions. In spite of this, little was actually known about the World Ocean as a coherent body and fragments of information failed to constitute actual knowledge about the ocean-space, which remained vastly unknown and ill understood as a whole. Parts did coalesce into a whole, or, following Poincaré's (1908, 168) analogy:

> We make science with facts just like we make a house with stones; but a heap of stones is no more a house than a collection of facts is science.

Things were known about the ocean-space, but they did not constitute scientific knowledge about the ocean-space as a geographical entity.

What the Enlightenment achieved, then, was to turn an assortment of facts into an organized, contained body of science. The factual was differentiated from the experienced and the phenomenological. What constitutes a fact was defined and factual knowledge was recognized as knowledge, which had been secured through stringent processes that evaluated sources, weeded out myth, and established reliable guidelines for scientific method. The objective was that knowledge should become a defensible, reliable, and permanent body of verifiable facts. As Priestley (in Kramnick 1995a, xiii) wrote, the Enlightenment saw science "overturn in a moment (. . .) the old building of error and superstition. [Science became] the means, under God, of extricating all error and prejudice, and of putting an end to all undue and usurped authority." Crucially, the Enlightenment's methods were applied to all aspects of human knowledge, thereby revolutionizing the entirety of the human experience of the earth.

With regard to geography and knowledge of the earth, the Enlightenment saw the globe drawn onto charts and maps: places were grid-referenced into coordinates, feelings mutated into numbers, superlatives turned into figures. As Harvey (in Livingstone and Withers 1999c, 13) writes:

> The Enlightenment Project (. . .) amounted to an extraordinary intellectual effort on the part of the Enlightenment thinkers to develop objective

science, universal morality and law, and autonomous art according to their inner logic. (. . .) [The] development of rational forms of social organization and rational modes of thought promised liberation from the irrationalities of myth, religion, superstition, release from the arbitrary use of power as well as from the dark side of our own human natures.

Effectively, the Enlightenment mathematicized the world and its representations, after establishing a scientific dogma on all facets of society. While certain aspects of it arguably gave way to postrevolutionary Europe, and Marxist and utilitarian philosophies, the Enlightenment succeeded in replacing primeval mythologies with a solid scientific grounding that still prevails philosophically, practically, and theoretically. As Kramnick (1995b, xii-xiii) writes:

> The eighteenth century began the Western love affair with science and technology that only now shows signs of being broken up by environmentalism and certain strands of postmodernism. Science embodied reason, and a scientific worldview embodied a rational perspective freed from religion and superstition.

Science replaced religious superstition and myth, as everything became "subject to a critique of reason if it were to command itself to the respect of humanity" (Kramnick 1995b, xi). This, as seen in the previous chapter, had a clear impact on the discovery of the Atlantic Ocean and the American continent as an entity and the global ocean-space more generally.

Yet while the goals and methods of the Enlightenment are, with hindsight, clear, like all historical movements, its time frame is more a matter of convention than being marked by actual, distinguishable events. The Enlightenment did not begin or end overnight; rather, it was conceived of retrospectively as a coherent set of ideas. While its spirit guided its dogmas, Withers (2007, 1) writes:

> for its contemporaries it was then, and for modern scholars it is now, an intellectual movement distinguished by critical, analytic, and scientific concerns and by claims that the power of reason could improve the human condition. Rather than being a fixed set of beliefs, the Enlightenment—as a moment and a movement—was a way of thinking critically in and about the world.

The Enlightenment as referred to today is a retrospective concept that historical periodization has allowed to demarcate. Yet for the convenience of working with the fixed boundaries that it offers, periodization "ranks among the more elusive tasks of historical scholarship" (Bentley 1996, 749). Bentley continues:

> The identification of coherent periods of history involves much more than the simple discovery of self-evident turning points in the past: it depends on prior decisions about the issues and processes that are most important for the

shaping of human societies, and it requires the establishment of criteria and principles that enable historians to sort through masses of information and recognize patterns of continuity and change.

(Bentley 1996, 749)

In spite of its inherent shortcomings, periodization remains a useful tool, and I will use it here to frame the current discussion and indeed define a period more relevant to this present study. This period will digress from conventional borders of the Enlightenment, but I will justify the used limits in relation to the specific discussion of understanding the ocean-space's geography.

The long eighteenth century, incorporating the latter part of the seventeenth century and the Age of Reason, and ending with the downfall of Napoleon Bonaparte and his exile to Saint Helena in 1815, is the period usually considered to be the Enlightenment. These limits have to do with the political contexts on a European scale of events, including, in particular, the French Revolution and its immediate aftermath. These particular limits, however, are uninteresting here inasmuch as they are principally land based and not primarily concerned with the scientific aspect of discovery unless it bears territorial claims; I will therefore reject them here so that the ocean-space can be better positioned at the center of geographical discourses and ways of understanding space. This is core to this book's argument that the ocean-space should not be relegated to the geographical sidelines or thought of only in opposition to land. To periodize the Enlightenment in a manner useful to this goal, focus will therefore be on defining a period of mathematical, geographical, and philosophical agreement in relation to the enterprise of knowledge making at sea and about the ocean-space. Significantly longer than the eighteenth century, this oceanic Enlightenment is concerned more directly with science and less by its human political and social matters.

For the current purposes, then, the Enlightenment will begin in 1637 with the publication of René Descartes' *Discourse on Method*. The subtitle of the book specifies that the method is to "guide one's reason well and search for truth in sciences" (Descartes 1991). This treatise for scientific method is, arguably, a kind of user guide for Enlightenment science and is also crucial in the epoch's cartographic advancements. Since it clearly articulates the scientific method and, in particular, the processes of knowledge production and study, it is suitable as the starting point of the Enlightenment, which, as will be made evident further, was so important to geographical exploration. The book indeed encapsulates the methodology of the Enlightenment that saw "the world as an object of geographical inquiry" and was concerned with "processes of discovery rather than with questions of scale" (Withers 2007, 87).

Two hundred and thirty-five years later, in 1872, Captain George Nares sailed from Portsmouth aboard HMS *Challenger*. For three years, Nares and a crew of scientists traveled the world's oceans, measuring its properties such as depth, surface movements and currents, and temperatures at different depths. Scientists also collected samples and logged the weather, and described the environments they sailed through. The scientists onboard collected as much data as possible to produce as much knowledge about the ocean. The *Report of The Scientific Results of the Exploring Voyage of HMS Challenger during the years 1873–76*, which was published over the decades following the ship's return to port, is a monument to modern science and scientific method (Thomson et al. 1880–1895). It details the most complete understanding of the ocean-space that had ever existed until then. With regard to ocean science and geographical knowledge of the ocean-space, the HMS *Challenger* expedition therefore epitomizes the spirit of the Enlightenment and the desire to know things scientifically. However, the technology used on the HMS *Challenger* expedition was very much grounded in the nineteenth century, which cements the expedition in a premodern era: for instance, the ship was a sailboat and only used its steam engine for maneuvers, the sounding equipment was rope rather than piano wire (Rice 2001). For these reasons, the return of HMS *Challenger* to port in 1876 seems suitable to mark the end of the oceanic Enlightenment.

Between 1637 and 1876, the oceanic Enlightenment frames the scientific thread of the period's geographical enterprises. However, as Foucault (in Livingstone and Withers 1999b, 19) writes:

> We must never forget that the Enlightenment is an event, or set of events and complex historical processes, that it is located at a certain point in the development of European societies. As such, it includes elements of social transformation, types of political institution, forms of knowledge, projects of rationalization of knowledge and practices, technological mutations that are very difficult to sum up in a word, even if many of those phenomena remain important today.

The scientific inquiry of the period is essentially inseparable from the social, humanist, and ideological shifts that were taking place at the same time; however, these will be useful here only inasmuch as they are linked with the period's science. Therefore, the borders of the oceanic Enlightenment here presented are scientific, from the invention of scientific method to a testing of this method and the creation of scientific knowledge. With this as a guideline, the oceanic Enlightenment concurs with the original Kantian definition as

> [Man's] release from his self-incurred tutelage. Tutelage is man's inability to make use of his understanding without direction from another. Self-incurred is this tutelage when its cause lies not in lack of reason but in lack of resolution to use if without direction from another. *Sapere aude!* "Have courage to use your own reason"—that is the motto of Enlightenment.
>
> (Kant 1995, 1)

Here the combination of reason and science can produce knowledge. This will be enlightened knowledge, and is the only kind of knowledge that matters in the current context. The oceanic Enlightenment can now be utilized to contextualize the upcoming debates and the understanding of the production of knowledge about the ocean-space.

Chapter 3 discussed the use of the ocean-space as a framework for historical and geographical study with specific material and spacial properties. This has also lead to dismissing national and local strands of the Enlightenment to concentrate on the larger debates at play. However, as the phrase "Atlantic Enlightenment" exists, it must be discussed. The phrase is seldom used and never with much agreement as to what it refers to. Indeed, Withers (2006b, 60) writes that the "Atlantic Enlightenment, if it has been considered at all, has achieved the status of an 'absent presence,' a taken-for-granted space." It emerges as "a scientific space of margins and of flows," a "space for knowledge's making and mobility between communities, [involving] issues of trust and of instrumentation to do with ships, sextants and logbooks, and of a social type of intellectual authority to do with mariners' knowledge" (Withers 2006b, 60 and 62). The phrase merges space, technology, and trust, encapsulating a located manner of "interpreting the Enlightenment geographically" but not necessarily addressing any specific issues (Withers 2006b, 61). Significantly, then, it may be more useful to think of the Atlantic Ocean with regard to the Enlightenment "not as a single entity but as a dynamic space, at once social, geographical and epistemological" whose "characteristics were different depending on where knowledge was made and received, by whom, in what form and in how such knowledge and its makers moves, if at all, between one place and another" (Withers 2006b, 62). Therefore, the Atlantic Enlightenment as examined here by Withers merely reads expressions of the Enlightenment as being played out on the Atlantic Ocean, bearing in mind various debates of spatial and technical dynamics. Conversely, the period of the oceanic Enlightenment that I defined earlier is concerned with the entwinements of science and elements of knowledge production that shaped the ocean-space scientifically. This latter outlook examines the ocean-space from a wider geographical and historical perspective than the phrase "Atlantic Enlightenment" does.

In addition, rather than being solely theoretical, the importance of the Enlightenment with regards to knowledge production of the ocean-space can be demonstrated through particular events, which contributed to the production of knowledge about the ocean-space. Such examples, which I will now turn to, are the making and unmaking of nonexistent islands in the Atlantic Ocean, the issue of coordinating geographical data across the globe, oceanic water movements across the globe, and the measuring of magnetic fields and their contribution to knowing the ocean-space. Original archival material from the United Kingdom Hydrographic Office (UKHO) will underpin particular stories and show how these come together as ways of producing geographical knowledge about the ocean-space.

4.2 Producing knowledge about the Atlantic ocean-space

4.2.1 *Making and unmaking nonexistent islands*

In 1525, somewhere in the Gulf of Guinea, on his way to the Maluku Islands in Indonesia, García Jofré Loyasa saw a landmass on the horizon. A few days previously, he had encountered a Portuguese ship whose crew had given him directions to a nearby, little-known island, which he now thought he was looking at. What follows is that, in October 1525, Loyasa set foot on

> a high land, four leagues around, covered with palms and oranges. There were no inhabitants, only human bones upon the ground, remains of houses, and a wooden cross.
>
> (Strommel 1984, 121)

As this island did not appear on any charts Loyasa had available to him, he took matters in his own hands and corrected the chart he was using by drawing, at approximately two degrees of latitude south and eight degrees of longitude west, the Island of Saint Matthew.[1] Despite a distinct lack of further evidence that Saint Matthew was not one of the already known and charted islands off the coast of Africa, the island continued to appear on maps and surveys until the nineteenth century. Among others, it appears on Battista Agnese's *Portolan Chart of Europe, Africa and the America*, c. 1550, *West Africa from le Neptune Francois* in 1693, and *Chart of South America and the Southern Ocean* in 1807.[2] These charts represented, each in their own time, the finest geographical knowledge available, yet, invariably, they were wrong, if only for their inclusion of the Island of Saint Matthew.

Confirming the existence and exact location of the Island of Saint Matthew was the aim of several naval expeditions in the period between being first created, though somewhat inadvertently, and being disproved. In 1817, HMS *Julia* went on a month-long expedition zigzagging in the zone in which Loyasa had allegedly seen the island. The journey is charted on *The Track of His Majesty's Sloop Ship Julia of 1817*.[3] On September 16, 1817, HMS *Julia* departed Saint Helena, made her way to, approximately, two degrees thirty minutes of latitude south and one degree of longitude west by 21 September. There she steered westward, meandering to the meridian of fourteen degrees longitude west, before turning back on 4 October, apparently seeing a rock on 15 October, and heading southward toward Ascension Island, which she reached on 21 October. At least two other ships, HMS *Inconstant* in 1817 and the *Cigogne* in 1833 also set out to find the island (Strommel 1984, 121). In each case, unsurprisingly in retrospect, no island was ever found in the vicinity of two degrees of latitude south and eight degrees of longitude west, and so the Island of Saint Matthew was eventually erased from maps and charts.

Similarly, the Isle Grande appears on the aforementioned 1807 *Chart of South America and the Southern Ocean* and the 1808 *Chart of the Ethiopic or Southern Ocean, and part of the Pacific Ocean*.[4] On the former of these two maps, the Isle Grande is positioned in three different places, all on the same line of latitude but a few degrees of longitude apart. Snaking around the supposed islands are the tracks of two ships: that of "Monsieur de la Pérouse" in 1785 and that of Captain Vancouver in 1795, both of whom were searching for the Isle Grande. The captions next to each location read thus: "I. Grande according to Mr. Dalrymple", "Probable situation of the I. Grande; but very uncertain" and "I. Grande. A good harbour. Discovered by la Roche in the year 1675 and as laid down by Capt. Cook's (*sic.*), but erroneously, as it appears by the track." The track referred to here is a section of M. La Pérouse's search for the island in 1785. On the 1808 map, the cartographer resorted to the same technique of drawing the island three times, annotating each to indicate the uncertainty of his claims. The Isle Grande locations are marked, again, to reflect the origin of the source thus: "I. Grande according to Dalrymple", "I. Grande. A good harbour. Discovered by la Roche in 1675. Situation very uncertain" and "I. Grande. According to Capt. Cooks (*sic.*) chart." The captions are thus hierarchized, leaving to others the choice of deciding which location to use. This method of organizing facts according to their source was by no means unique, as I return to discuss at a later stage in this chapter.

Further south, the Auroras Islands, situated halfway between the South Georgia Islands and the Falkland Islands, also have a complicated history of nonexistence. Most notably, these

> have the distinction of having been precisely surveyed by a genuine Spanish naval hydrographic vessel, the corvette *Atrevida*, which had been detached, in 1774, from a general survey of the Patagonian coast under the overall command of Allessandro Malaspina. This survey was part of a two-ship scientific cruise around the world, well enough respected by geographers to earn Malaspina the honor of having his name bestowed on a glacier in Alaska and a volcano in the Philippines. (. . .) The Auroras were reported at first by the *Aurora* and then by the *Princess*. To the Spanish authorities, their position was a danger for the trade between Lima and Cadiz; a survey was called for. The *Atrevida*'s survey was very precise. She carried chronometers; and had checked them only a few days earlier at Soledad in the Falklands. But the Auroras did not exist.
>
> (Strommel 1984, 84)

Nonetheless, the Auroras duly appear on the 1808 *Chart of the Ethiopic or Southern Ocean and part of the Pacific Ocean* where Malaspina saw and surveyed them. Until the latter half of the nineteenth century, the Auroras were still drawn on maps and official accounts not only in Spain but also in Britain. However, failure to find them again after Malaspina's survey rapidly suggested that they did not exist.

These three nonexistent islands are but a few from a collection of many that appeared on maps of the Atlantic Ocean while it was being explored, discovered, and mapped. Among others are the Pepys Islands, north of the Falkland Islands, which were searched for by the ship *Tamar* in 1808 (see *Carte Esferic de Oceano Meridional*[5]), Diego Alvarez Island, between Tristan da Cunha and Gough Island on the 1808 *Chart of the Ethiopic or Southern Ocean and part of the Pacific Ocean* in 1808, and Saxenbourg Island, also nearby. Importantly, this phenomenon was not limited to the Atlantic Ocean: fictional islands were also interspersed in the Pacific and Indian Oceans. Each event, wherever it may be, of by which islands came to be drawn on maps, then searched for and then, eventually, deleted, brings to the fore a specific process of making knowledge and acquiring facts about the ocean-space.

Indeed, over the years and with evermore technology, it has been possible to identify most of these phantom islands that populated the vast oceanic expanse as something else. In most cases, they were in fact cases of mistaken identity. Loyasa's Saint Matthew's island was probably São Tomé or Príncipe. The Auroras might have been one of the Falkland Islands, as was Pepys Island, and Diego Alvarez Island is likely to have been Gough Island. Other islands may have been volcanic and have now disappeared, while others could have been nothing more than optical illusions (which are prevalent at sea). However, these mistakes were only the side effects of Europe trying to understand the newly discovered Atlantic Ocean and fill its geographical emptiness. As such, the mapping of nonexistent islands emerges as part of the process of discovering facts about the ocean-space and an enterprise, which required specific skills and new technologies. Indeed, beyond the cosmographic shock and philosophical upheaval triggered by the discoveries of the Americas and the Atlantic and Pacific Ocean as seen in Chapter 3, the exploration of the ocean was a huge technological and scientific challenge. The technical difficulty of studying the ocean-space, and especially such challenges as the so-called longitude problem, the discovery of currents and large-scale water movements, and the mapping of earth's magnetism are three examples that highlight further challenges of knowledge production about the ocean-space. While these matters are not limited to the Atlantic Ocean, it is there that they were principally researched by Europeans, if simply for questions of proximity. Through considering these particular scientific stories and setting them against the backdrop of knowing the Atlantic Ocean, I continue to highlight how geographical knowledge about the ocean-space was formed in a complex interplay of a variety of factors.

4.2.2 Making coordinates

As well as physical discoveries of faraway lands being made during the Age of Exploration, two main obstacles pertaining to long-distance seafaring were overcome during its span: these are how to easily calculate longitude at sea and how to trace a ship's course on a plane surface. The latter had been in part solved by Gerardus Mercator with the invention in 1569 of the eponymous Mercator projection, which squared the globe and made it possible to trace courses on maps, while respecting the general geography of the route. However, the problem of longitude was a matter that was not successfully tackled until much later, considerably delaying any significant advance in knowing and charting the ocean-space and indeed the earth in general. Indeed, until it was possible to accurately locate ships at sea and, not insignificantly, shores encountered, it was impossible to create an accurate map of the ocean. It was not only the oceanic expanse that was unknown and unplaceable, so too were its actual limits and the lay of the land surrounding it. Further, without the ability to locate oneself, it became impossible to avoid even known dangers, essentially rendering some known facts unusable. This gave way to such situations that "during the eighteenth century, the Royal Navy was losing more ships due to running aground and floundering on rocks than to the actions of the enemy" (Holmes 2005, 266). Thus what was desperately needed by explorers and scientists alike was a method to rapidly and reliably calculate longitude at sea. This, it emerged, was a problem of scientific, technical, and cultural magnitude, which, when solved, provided a method to make geographical coordinates part of knowledge production about the ocean-space.

Before moving on, it is necessary to reintroduce the geographical coordinates positioning system that webs the earth in a geometrical manner and provides a framework to singularly identify points on its surface. This is to locate the discussion within a larger context of geographical knowledge. First developed by the Babylonians, but expanded by the Greek astronomer Ptolemy, the latitude and longitude system is a mathematical method that grids the surface of the globe with a series of imaginary lines, which create a graticule, in order to give each point a set of references that uniquely locate it. The Enlightenment saw a revival in the use of the graticule and arguably one "could not conceive of maps without graticules. The graticule was absolutely necessary for each map; it alone established the epistemological truth of geographic representation" (Edney 1999, 172). All lines of latitude are parallel to the Equator, and therefore to each other, and are perpendicular to the earth's axis of rotation. The number in a latitudinal reference stands for the angle in degrees formed between the equatorial plane, being a plane that passes through all the points on the Equator, and a vector that is normal to the location being studied. As the angle increases, the circumference of the circle decreases. On the other hand, a line of longitude is any imaginary line that joins the geographic North Pole to the geographic South Pole following a theoretical surface of the globe, which excludes local relief (that is if the earth were perfectly spherical). The number in a longitudinal reference stands for the angle in degrees between the plane of the referential Greenwich meridian great circle and a vector that is normal to the location being studied. All lines of longitude are of equal circumference and all pass through both the North and South Poles.

As well as being central to the creating of a mathematical model of the earth and, perhaps more importantly, providing a way of locating places in relation to one another, webbing the globe with imaginary lines is also essential to the needs of cartography. Cartography seeks to provide usable and accurate representations of the (spherical) earth on a two-dimensional surface. Establishing a globe-to-paper transference system is what map projections are principally concerned with. One example of these is the Mercator projection, which distorts size but retains shapes; another is the Gall–Peters projection, which distorts shape but preserves size. Combined with the graticule, projections allow cartographers to correctly place locales relative to one another but also within the bigger picture. The need for a reliable method to measure longitude and latitude in any given place at any given time therefore emerges as central to the enterprise of discovering and mapping the earth and its oceans.

Measuring the latitude of any geographical point is not a complicated process. In its simplest form, the method consists of measuring the angle between one's ship and two fixed points: Polaris (commonly known as the North Star) and the point on the horizon directly beneath Polaris. The angle measured is equivalent, to a near enough degree, to latitude of the ship. In the southern hemisphere where Polaris is not visible, constellations are used to locate the point in the sky above the South Pole, and that point is used to conclude degrees of latitude south of the Equator. During hours of daylight, a similar technique could be used by measuring the height of the sun at its zenith. In all events, what is important to highlight here is that the methods used to measure latitude were astronomical: positions were determined thanks to stars and with instruments designed by astronomers. With the ability to measure latitude, ships at sea could at least be located on a line. But narrowing down that position any more was impossible without the ability to measure longitude. For sure, the uses for latitude are severely limited if not combined with longitude.

Bearing in mind the ways of measuring latitude, the scientific consensus from the sixteenth century to the 1800s was that the method to measure longitude would be similarly astronomical. The Paris and London observatories were built with the prospect that they would be central to the ultimate discovery of the solution to the longitude problem. This would take time to be disproved and the rejection of the astronomical solution was, not unexpectedly, vehemently resisted by astronomers. Indeed, as Withers (2007, 97) writes, "Longitude's solution is less an unproblematic narrative of scientific advance than a complex tale of social competition and local geographical credibility." Without delving too deeply into the story of the invention of the chronometer, which was the surprising solution to the problem of measuring longitude at sea, I will nonetheless outline it briefly, because it is tightly entwined with maps that I return to shortly.

Astronomers who pored over the longitude dilemma include some of the most famous names of the period: Galileo Galilei, Jean Dominique Cassini, Christiaan Huygens, Isaac Newton, and Edmund Halley (Sobel 2007, 7). Two of the most resilient propositions put forward involved the movement of the moon relative to the stars in the sky and a method by Galileo using the moons of Jupiter. The former proved too complicated to implement as the movement of stars was still unclear, and the latter was very impractical at sea and useless during the day. Sobel (2007, 89) writes:

> A seaman could not read the clock of heaven with a quick glance but only with complex observing instruments, with combinations of sightings taken together and repeated as many as seven times in a row for accuracy's sake, and with logarithm tables compiled far in advance by human computers for the convenience of sailors on long voyages. It took about four hours to calculate the time from the heavenly dial—when the weather was clear, that is. If clouds appeared, the clock hid behind it.

However, Galileo's technique proved very useful and reliable on land where the right conditions could be awaited and the correct instruments well calibrated. Using it, it was possible to adjust maps of Western Europe, which had previously tended to inflate territories and kingdoms, perhaps in a combination of wishful regal thinking, imprecise instruments, and uncoordinated methods. When corrected, countries usually turned out to be, in fact, smaller than once thought. Thanks to Galileo's method, maps came to be based on facts extracted through tried and tested reliable methods, sometimes with unwelcome political consequences. In France,

> The picture of France found itself strangely shrunk. Louis XIV, to whom was presented the results, would have then joked, accusing the members of the Académie, of having subtracted a larger part of his kingdom than the most dishonorable war would have.
>
> (Dawson 2000, 15)

Indeed, to place this in context,

> The XVIIIth century began in what Roland Vere Tooley calls "one of the most brilliant periods in the history of cartography in France." The important works of the Royal Academy of Sciences, undertaken in the time of Colbert, provided the basis necessary for a reform of geography. Thanks to this multiplication of recognized astronomical positions, the cartographic drawing offices set up in the kingdom of the House of Bourbon based their science on solid, immovable ground. Until the end of the Ancien Régime, they influenced the whole of Europe in their cartographic conception, method and philosophy.
>
> (Dawson 2000, 15)

If only to make maps accurate, which is by no means a small feat, the astronomical method of calculating longitude was an important achievement and has been called "the cartographic equivalent of the *Encyclopédie*" (Withers 2007, 102). Yet, while astronomic mapping was undoubtedly useful, it did not provide the required "Practicable and Useful" manner of calculating longitude and definitely not so at sea (Longitude Act in Sobel 2007, 8).

When Parliament intervened to encourage, support, and set up a framework to test new methods of finding longitude at sea, the last thing that was expected was for the pocket watch to prevail over the celestial clock. However, this is what happened, and John Harrison's H4 chronometer emerged as a winner in a vicious competition. In short, H4 did not have a pendulum (which requires an immobile environment), so it could be used at sea (a very mobile environment). H4 did not suffer from barometric pressures and resisted humidity; therefore, it did not lose time while being rewound. In accordance with the terms of the contest, H4 was taken to Jamaica and back and without losing more than two minutes. All in all, H4 offered an easy solution to the problem of measuring longitude easily, being indeed "Practicable and Useful."

Nonetheless, once the issue of measuring longitude at sea was resolved, a further problem remained: How was the chronometer to be made easily accessible and affordable? Huge amounts of manpower had gone into inventing the device, but the expensive materials needed and the skill necessary to build chronometers made the device exclusive and difficult to manufacture in large quantities. When a London watchmaker called Larcum Kendall was commissioned to copy Harrison's watch, it took him two years and the Board of Longitude paid him £500 for his troubles (Sobel 2007, 153). This was a lot of time and effort for one clock, and he told the board as much, saying that "it would be many years (if ever) before a watch of the same kind with that of Mr. Harrison's could be afforded for £200" (Kendall in Sobel 2007, 153). Adding to the problem was the fact that sextants had in the meantime been invented and were now cheap. Combined with reliable charts and astronomical almanacs (ephemerides) they allowed a mariner to calculate longitude for less than £20, even though there was a great degree of human error possible with this method.[6] Eventually, it was watchmaker Thomas Earnshaw who both simplified the design of the chronometer and set up a production chain that could cheaply manufacture a marine chronometer within two months. It is thus that the chronometer ultimately became the object of reference to find out, rapidly and reliably, one's longitude anywhere on earth. By 1815, just eighty years after Harrison's marine timekeeper H1 first kept near-accurate time, approximately five thousand watches based on H4 were in existence, sailing the seas of the world (Sobel 2007, 163).

With regard to knowledge production about the ocean-space, the importance of the marine timekeeper is crucial in that it is tightly entwined with the ability to measure and represent the ocean-space accurately. Certainly, the mapping of the ocean-space could claim to be accurate only if its space could be mapped within a relative, determined graticule. Thus solving the issue of longitude allowed for maps and charts that had been drawn previously to be corrected. The example of France is just one of many: cartographers in most European nations used the marine chronometer to correct their nations' maps. Scotland, Prussia, Saxony, the Hapsburg's lands and Britain followed, before also remapping their respective, far-off colonies (Withers 2007, 103–104). As it became possible to correctly locate islands, there were many instances, as the ones shown earlier, of islands moving to reflect different longitudinal readings. Another example can be found on two maps of Havana, one Spanish and the other its British copy. The first is *Plano del Puerto y Ciudad de La Havana*[7] and the second one is its British *Plan of the Harbour and City of the Havana*.[8] On the Spanish chart, the referential line of longitude goes through the Moro Fort and is marked 76° 00' 36" west of Cadiz. On the British map, the same line is named in reference to the Greenwich meridian, indicating the same fort at 82° 15' 36" west of Greenwich. This gives the longitudinal difference between Cadiz and Greenwich to be six degrees and fifteen minutes. While they are among the best maps of their times, all of these measurements are inaccurate and were eventually easily corrected. Indeed, the city of Havana lies eighty-five degrees, twenty-one minutes, and twenty-five seconds west of Greenwich, signifying that the surveys were off by six minutes and eleven seconds. Further, Cadiz is six degrees and eighteen minutes west of Greenwich. While these differences in degrees do not translate into huge distances on the ground (approximately five miles), they are nonetheless incorrect by any scientific standard. The timekeeping device, albeit together with better astronomical techniques, allowed for places to be accurately placed and maps corrected, highlighting how, as Withers (2007, 97) writes, "Enlightenment travel involved bringing the world to light less by imposing a single universal standard—than by calibrating others' local standards with a view to ensuring, in time, a commensurability over space. (. . .) [L]ocation makes a difference to the production of knowledge and to the context of its reception and justification." Certainly here, the manner in which knowledge is located in relation to a graticule is essential in order to know the ocean-space in a geographically coherent manner.

Furthermore, and equally primordial to navigation, the use of the chronometer to instantly position oneself at sea was a significant step toward creating knowledge about the ocean-space while navigating its moveable surface. Indeed, the chronometer enabled in situ observations with immediate readings to be taken. Combined with the use of increasingly accurate maps, this was a huge progress in navigation. Instead of using the very uncertain method of dead reckoning[9] or very limiting method of latitude navigation,[10] the chronometer deleted a huge amount of uncertainty in the drawing of ships' courses on maps. What began with the advent of the chronometer was the appearance of comparative routes on maps: one traced route would represent the position by dead reckoning and the other the route traced by chronometer readings. For instance, the 1805 with additions in 1812 *Chart of the Atlantic Ocean* shows ships' courses by dead reckoning as well as those corrected by use of the chronometer.[11] Gradually, as the use of the technique of finding longitude by use of the chronometer was being generalized, charts were being corrected and navigation made more precise.

A New General Chart of the Atlantic or Western Ocean and Adjacent Seas of 1803[12] shows several ways of marking positions. Some are marked clearly with a star as being determined astronomically, others as "Latitude only is known" are showed with a cross, "longitude has been determined by the help of the marine clocks of Mr. Berthoud" is marked with a dot in a circle and "Latitude has been observed at sea with the octant by Mr. de Fleurieu" is indicated with a double cross. These symbols, branding the origins of knowledge, organize facts according to the method that was used to acquire them. The map user is thus invited to rate or hierarchize the facts and decide which position is most trustworthy. This is similar to how Isle Grande was annotated on the 1808 *Chart of the Ethiopic or Southern Ocean and part of the Pacific Ocean* discussed earlier. In time, however, the chronometer would phase out astronomical longitude calculations, changing the standards of accuracy and creating a new version of facts. With the chronometer, the latitude and longitude grid was cemented, having gained credibility through testing finally validated by the Board of Longitude and experience. The globe was now webbed by verifiable lines, giving credence to the system, allowing for the earth to be known rather than estimated and guessed, and for the shape of the ocean-space and the continental masses to be discovered. The chronometer also cleared the oceans of nonexistent islands that had made their way to nautical charts because of poorly estimated longitudes. With Harrison's timekeeper, a navigator could establish longitude from a map table quite rapidly, and mapmakers on shore could claim that their new charts beheld a new kind of knowledge: one that was verifiable and mathematically (if not yet politically) standardized. The chronometer demonstrably changed the manner in which geographical knowledge was produced and the way in which the ocean-space became measured relative to the rest of the earth, representing an instance of production of geographical knowledge exemplary of the Enlightenment knowledge shift. Another such example will be discussed now, examining large-scale water movements.

4.2.3 Water movements across the earth

In the first half of the nineteenth century, Matthew Fontaine Maury began working on *Physical Geography of the Seas and Its Meteorology* (Maury 1855). Within a few years of its first publication, the book had reformed the way transoceanic trips were planned, as ships began using currents to their advantage instead of being held hostage by ill-understood, uncharted water movements. With regard to the making of knowledge, though, what is further worth noting is the source of inspiration from whence Maury's work derived: what spurred Maury's theory and research was a biblical psalm. Psalm 8 describes the beauty of the earth made by God and ponders that what he has given man. The psalm reads:

> [God] made [Man] ruler over the works of [God's] hands; [he] put everything under his (*sic*) feet: all the flocks and herds, and the beasts of the field, the birds of the air, and the fish of the sea, and all that swims the paths of the seas.
> (Psalm 8: 6–8)

For Maury, this was an irrefutable statement that the ocean had pathways of one sort or other and that all that needed to be done was to discover their nature and locate them. In this sense, Maury sought out to find something he knew was there, rather than discover something new: the existence of "paths of the seas" was not questionable, only their nature and location. In all events, by 1860, Maury's work had contributed to drastically changing the nature of sailing the world's oceans as accumulated knowledge allowed for sailing routes to be shortened and perfected. In practice, Maury's approach follows Cornish botanist Jonathan Couch's assessment that science "is the knowledge of the works of God [which] can only be known by personal examination" (Couch in Naylor 2005b, 37). Certainly, in the nineteenth century, religion was still a strong cultural influence, thus seeping into scientific discourse and influencing the making of scientific knowledge, which remains, following Livingstone (2003), tightly entwined within the settings of the period within which it is produced. However, the place of religion was no longer at the center of geographical discussions whence, as seen earlier, it had begun to be superseded in the fifteenth century.

Before Maury delved into the mysteries of large-scale water movement, Benjamin Franklin had already suggested that something was drifting in North Atlantic waters. The Gulf Stream, while not entirely unknown, was, however, neither accurately located nor had ever been charted. In a long letter to Mr. Alphonsus le Roy in Paris, Franklin was the first to draw a picture of the elusive current. This letter touches on a variety of subjects pertaining to navigation, but I will only examine its section on the Gulf Stream in detail here.[13]

Equal to the magnitude and importance of the map that Franklin included is the deductive process that led to Franklin's assertive drawing. For this reason, I have quoted Franklin's paragraph "Sundry circumstances relating to the Gulph Stream" in its entirety in order to emphasize two things: the original thought process and the importance of the chart with regard to its direct applications to the daily passages of the many ships sailing between Europe and North America.[14] With regard to this book, these two characteristics are essential to bring to the fore the processes of knowledge production on the ocean-space. The passage reads thus:

> Vessels are sometimes retarded, and sometimes forwarded in their voyages, by currents at sea, which are often not perceived. About the year 1769 or 70, there was an application made by the board of customs at Boston, to the lords of the treasury in London, complaining that the packets between Falmouth and New York, were generally a fortnight longer in their passages, than merchant ships from London to Rhode Island, and proposing that for the future they should be ordered to Rhode-Island instead of New York. Being then concerned in the management of the American post office, I happened to be consulted on the occasion; and it appearing strange to me that there should be such a difference between two places scarce a day's run asunder, especially when the merchant ships are generally deeper laden, and more weakly manned than the packets, and had from London the whole length of the river and channel to run before they left the land of England, while the packets had only to go from Falmouth, I could not but think the fact misunderstood or misrepresented.
>
> There happened then to be in London, a Nantucket sea-captain of my acquaintance, to whom I communicated the affair. He told me he believed the fact might be true; but the difference was owing to this, that the Rhode-Island captains were acquainted with the Gulf Stream, which those of the English packets were not. We are well acquainted with that stream, says he, because in our pursuit of whales, which keep near the sides of it, but are not to be met within it, we run down along the sides, and frequently cross it to change our side: and in crossing it have sometimes met and spoke with those packets, who were in the middle of it, and stemming it. We have informed them that stemming a current, that was against them to the value of three miles an hour; and advised them to cross it and get out of it; but they were too wise to be counselled by simple American fishermen.
>
> When the winds are but light, he added, they are carried back by the current more than they are forwarded by the wind: and if the wind be good, the subtraction of 70 miles a day from their course is of some importance. I then observed that it was a pity no notice was taken of this current upon the charts, and requested him to mark it out for me which he readily complied with, adding directions for avoiding it in sailing from Europe to North America. I procured it to be engraved by order from the general post office, on the old chart of the Atlantic, at Mount and page's, Tower hill, and copies were sent down to Falmouth for the captains of the packets, who slighted it however; but it is since printed in France, of which edition I hereto annex a copy.

This stream is probably generated by the great accumulation of water on the eastern coast of America between the tropics, by the trade winds which constantly blow there. It is known that a large piece of water ten miles broad and generally only three feet deep, has by a strong wind had its waters driven to one side and sustained so as to become six feet deep, while the windward side was laid dry. This may give some idea of the quantity heaped up on the American coast, and the reason of its running down in a strong current through the islands into the Bay of Mexico, and from thence issuing through the Gulph of Florida, and proceeding along the coast to the banks of Newfoundland, where it turns off towards and runs down through the Western Islands.

Having since crossed this stream several times in passing between America and Europe, I have been attentive to sundry circumstances relating to it, by which to know when one is in it; and besides the gulph weed with which it is interspersed, I find that it is always warmer than the sea on each side of it, and that it does not sparkle in the night: I annex hereto the observations made with the thermometer in two voyages, and possibly may add a third. It will appear from them, that the thermometer may be an useful instrument to a navigator, since currents coming from the northward into southern seas, will probably be found colder than the water of those seas, as the currents from southern seas into northern are found warmer.

And it is not to be wondered that so vast a body of deep warm water, several leagues wide, coming from between the tropics and issuing out of the gulph into the northern seas, should retain its warmth longer than the twenty or thirty days required to its passing the banks of Newfoundland. The quantity is too great, and it is too deep to be suddenly cooled by passing under a cooler air.

The air immediately over it, however, may receive so much warmth from it as to be rarified and rise, being rendered lighter than the air on each side of the stream; hence those airs must flow in to supply the place of the rising warm air, and meeting with each other, form those tornados and waterspouts frequently met with, and seen near and over the stream; and as the vapour from a cup of tea in a warm room, and the breath of an animal in the same room, are hardly visible, but become sensible immediately when out in the cold air, so the vapour from the Gulph Stream, in warm latitudes is scarcely visible, but when it comes into the cool air from Newfoundland, it is condensed into the fogs, for which those parts are so remarkable. The power of wind to raise water above its common level in the sea, is known to us in America, by the high tides occasioned in all our sea-ports when a strong northeaster blows against the Gulph Stream.

The conclusion from these remarks is, that a vessel from Europe to North America may shorten her passage by avoiding to stem the stream in which the thermometer will be very useful; and a vessel from America to Europe may do the same by the same means of keeping in it. It may have often happened accidentally, that voyages have been shortened by these circumstances. It is well to have the command of them.

(Franklin 1806, 185–188)

This passage is crucial for several reasons. It demonstrates a methodology that is, while unconventional in its use of whales as a scientific instrument, strict and framed within scientific logic. Several sources are used and cross-checked, measurements are taken, theorems enumerated. The object of Franklin's efforts are to turn hearsay about the Gulf Stream into irrefutable fact, even if certain details, admittedly, still need honing. As Withers writes:

> there was, of course, nothing wrong with relying on facts borne of experience. But for those crossing the world's oceans or mapping their borders, accepting others' practical knowledge—in this case that of ships' captains and common seamen—meant treating as equal people who, in other circumstances, would not be deemed legitimate sources of rational knowledge.
>
> (Withers 2007, 131)

However, Franklin uses scientific method to try and test the affirmations of "simple American fishermen."

Indeed, there is a tension regarding the question of what constitutes knowledge between two groups who claim to know. Franklin's Nantucket acquaintance knew

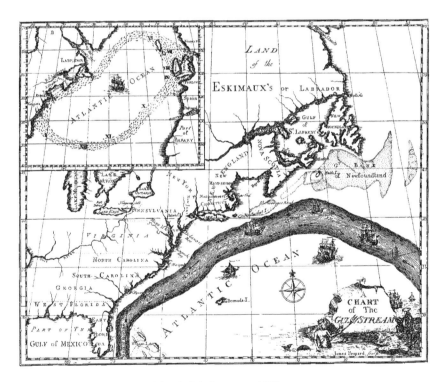

Figure 4.1 Franklin's chart of the Gulph Stream, c. 1770.

Source: http://commons.wikimedia.org/wiki/File:Franklingulfstream.jpg

of the Gulf Stream and had been using it on his whaling expeditions. While the sea captain's way of knowing was not strictly scientific and was solely based on observation of what worked on any given day, the captain still knew that waters were adrift. On the other hand, "those on English packets" have no reason to know of the stream and have therefore been delayed by it as it forced them to tread water in countercurrents rather than steadily advance toward the New World. Yet, when informed of the current, the English preferred to ignore these facts that came from what they perceived as an unreliable source, the fishermen. In effect, what was happening between the fishing vessels and the navy ships stemming the Gulf Stream is reminiscent of what I have explored in the previous chapter concerning the mental discovery of places. In effect, the refusal to acknowledge something new until it is backed up by several, accepted, testable, and reliable methods follows the method outlined by Descartes to know scientific truths. The fishermen lacked authority in the eyes of the British, and thus knowledge was lost in the convoluted corridors of protocol. Until it gained weight and was eventually added to British Admiralty maps, knowledge of the Gulf Stream was questionable. Following Livingstone (2003), the nature of knowledge is here shown to be tightly entwined with the nature of authority whence it comes and raises the question of what makes a source credible. In this case, what came first was the credibility of the phenomenon, as charts and surveys began showing the "probable course of the Gulf Stream" as an arced line across the North Atlantic, such as the 1805 with additions in 1812 *Chart of the Atlantic Ocean*, while in the meantime, others still found it necessary to single out "intelligent fishermen" or a "mere whaler" when they were encountered (Couch in Naylor 2005b, 39; Bravo 2006, 516). This leads to Withers (2007, 60–61) questions about the places of scientific knowledge, writing:

> For Franklin, the question "how and where was oceanographic and navigational knowledge made in the Enlightenment?" would allow different answers: at sea, as ships crossed the ocean and one another's path, even if English captains refused to acknowledge the experiential knowledge of their American counterparts whose navigational practices were based on natural observation—whale watching; on a map, produced in a London tavern as a result of the verbal testimony of a Nantucket sailor (and allowing for the fact that the map in question was later slighted by its intended audience); in Paris as a letter received; and in Philadelphia as an article read by fellow members opening their *Transactions*.

Here, the networks of credibility and the geographies of trust are brought to the fore, precisely locating the making of knowledge within a complex web of interactions. As such, the Enlightenment emerges "as a period of intellectual mobility when 'a general transformation of different kinds of knowledge into different forms of knowledge took place'" (Withers 2007, 61).

Furthermore, Franklin's letter is remarkable in that it explains how this new knowledge can be used easily in order to simplify navigation and make routes more trustworthy. Franklin appears to have anticipated the need for his formalization of the Gulf Stream to have a practical application, lest it remain uncared for. In response to the initial query submitted to the post office, Franklin saw a direct application to knowledge of the Gulf Stream: facilitating ships' crossings. By taking advantage of currents, journey lengths could be shortened and arrival dates be more predictable. Comparatively to what Maury would achieve by the mid-nineteenth century, Franklin's suggestions are very limited; yet, they represent the movement toward using knowledge for specific ends.

Returning to Maury, his undertakings and the scope of the results of his efforts to combine dozens of ships' logs so that a single chart could "show [a young navigator] not only the tracks of the vessels, but the experience also of each master as to the winds and currents by the way, the temperature of the ocean, and the variation of the needle" is difficult to conceptualize (Maury 2003, 3). However, one way of assessing the impact of Maury's charts is by the rapid acceptance of the new data by sailors, and the number of days at sea avoided by using the sea's movements rather than simply riding it. Maury (2003, 3–4) writes:

> Such a chart could not fail to commend itself to intelligent shipmasters, and such a chart was constructed for them. They took it to sea, they tried it, and to their surprise and delight they found that, with the knowledge it afforded, the remote corners of the earth were brought together, in some instances, by many days' sail. Before the commencement of their undertaking, the average passage to California was 183 days; but with these charts for their guide, navigators have reduced that average, and brought it down to 135 days.

Between England and Australia, the average time going without these charts, is ascertained to be 124 days, and coming about the same; making the round voyage one of about 250 days on the average

> These charts, and the system of research to which they have given rise, bid fair to bring that colony and the mother country nearer by many days, reducing, in no small measure, the average duration of the round voyage.

Furthermore, because of the cost of each day at sea, it is possible to put monetary values to the economies ensured by virtue of shortened trips. This was foremost in the mind of merchants, and calculations were made to transform Maury's charts into savings. This highlights the relationship between knowledge of the earth and commercial ventures, the former often being a consequence of the latter, or, as Withers emphasizes, "encountering the world empirically in an age of European empires meant that trade and learning went hand in hand, but they did not do so equally" (Withers 2007, 87). Certainly, following Burstyn (1975), the economic scale is helpful in order to assess the practical applications of scientific endeavors. Further, making sense of the value of science in relation to economic factors bears witness to the close links between science and how it might be applied practically, thus showing once more the tight entwinement between various areas of socio-cultural settings that underpin knowledge production. However, the economics of commerce aided by the increasing production of knowledge does not fall directly within the goals of this book. Nonetheless, I include this brief discussion here as, in this case, economy provided a good way of understanding Maury's work without hijacking its scientific aims.

Hunt (1854, 546–547) calculated the savings enabled by Maury's work as follows:

> Now let us make a calculation of the annual saving to the Commerce of the United States effected by those charts and sailing directions. According to Mr. Maury, the average freight from the United States to Rio de Janeiro is 17.7 cts. per ton per day, to Australia, 20 cents, to California also about 20 cts. The mean of this is a little over 19 cents per ton per day, but to be within the mark, we will take it at 15, and include all the ports in South America, China and the East Indies.

The sailing directions have shortened the passages to California to thirty days, to Australia to twenty, to Rio Janeiro to ten. The mean of this is twenty, but we will take it at fifteen, and also include the aforementioned ports in South America, China, and the East Indies.

We estimate the tonnage of the United States engages in trade with these places at 1,000,000 tons per annum. With these data, we see that there has been effected a saving for each one of these tons of 15 cents per day, for a period of 15 days, which will give an aggregate of $2,250,000 saved per annum. This is on the outward voyage alone, and the tonnage trading with all other parts of the world is also left out of calculation. Take these into consideration, and also the fact that there is a vast amount of foreign tonnage trading between these places and the United States, and it will be seen that the annual sum saved will swell to an enormous amount.

By studying how the ocean moves, Maury effectively succeeded in having a tangible effect upon world trade, accelerating human relations, and the transit of goods, as well as economizing millions of dollars in the process. It is the scientific process of Maury's work, however, which is most relevant to the production of geographical knowledge. This involved logbooks buried deep in forgotten chests brought out in an international effort to make science and advance knowledge of the seas.

In many respects, the groundwork of Maury's project was initially one of massive coordination as "a ransacking of time-honored sea-chests in all the maritime communities of the country for old log-books and sea journals" was meticulously and laboriously turned into track charts that "divided the ocean off into great turnpike-looking thoroughfares" (Maury 1855, v; Maury 1855, ix). The route to China, the great highways to Brazil, and the passages to Australia were revealed as a massive effort to compile data brought together the maritime community. Maury (2003, 4) wrote:

> A system of philosophical research which is so rich with fruits and abundant with promise could not fail to attract the attention and commend itself to the consideration of the seafaring community of the whole civilized world. It was founded on observation; it was the result of the experience of many observant men, now brought together for the first time and patiently discussed. The results tended to increase human knowledge with regard to the sea and its wonders, and therefore they could not be wanting in attraction to right-minded men.

If only those whose livelihoods are the sea found the new charts useful on a daily basis, the advancement of knowledge was philosophically beneficial to all. In Maury's own words, "whatever the reapers should gather, or the merest gleaner collect, was to inure to the benefit of commerce and navigation—the increase of knowledge—the good of all" (Maury 1855, xii). And indeed, Maury's book was well received. Von Humboldt (in Hunt 1854, 537) wrote:

> I beg you to express to Lieut. Maury, the author of the beautiful charts of the Winds and currents, prepared with so much care and profound learning, my hearty gratitude and esteem. It is a great undertaking, equally important to the practical navigator and for the advance of meteorology in general. The shortening of the voyage from the United States to the Equator is a beautiful result of this undertaking.

Following the Kantian spirit of Enlightenment, Maury emerges here as advocating knowledge for knowledge's sake, the sources of which were reliable, if sparse and sporadic, and the compilation of which was often experimental and usually painstaking.

However, this twice-removed way of making science was also supplemented by measurements made of makeshift instruments. One particular method to try to, if not understand, at least visualize and map surface currents, was for mariners to throw a message in a bottle overboard and wait for it to wash up on a shore somewhere. The message consisted of the date and position at which the bottle was sent off. The idea was that when a bottle reached land, it would have done so following the "paths in the sea." Taking into account the lapse of time between launching and beaching, lines were drawn to estimate the route of the bottles and the directions to the seas' pathways.[15] It is thus that Captain Beechy of the Royal Navy charted the path of more than one hundred bottles. The results of this charting showed that,

> waters from every quarter of the Atlantic tend toward the Gulf of Mexico and its stream. Bottles cast into the sea midway between the Old and the New Worlds, near the coast of Europe, Africa and America, and the extreme north

or farthest south, have been found either in the West Indies, or within the well-known range of Gulf Stream waters.

(Maury 1855, 28–29)

Overall, these two uses in mapping large-scale water movements highlight that it was not a straightforward process but that it was possible to achieve with a certain degree of rigor. For this, Maury's compilation charts of paths in the sea are mind-blowing both for their range and their accuracy.

Maury's work on currents was not limited to their exploitation in terms of navigation but was driven by scientific and philosophical motors that sought to dissect the ocean in order to know it better. It was further intertwined with matters of meteorology and atmospheric movements, thus emerging as a book with more than just seafaring ramifications. Instead, it affects 70 percent of the earth's surface and all of its weather. Alongside *The Physical Geography of the Sea and its Meteorology*, Maury's 1851 *Explanations and Sailing Directions to Accompany the Wind and Current Charts* was among the first volumes to consider the earth as a whole whose elemental parts interacted so closely that it was impossible to consider them in isolation. (This will be discussed in further detail in Chapter 5.)

Von Humboldt epitomized this talking of the importance of studying meteorology from a global perspective, as well as the considerations of the earth as a whole. He writes:

> In the present condition of the surface of our planet, the area of the solid is to that of the fluid parts as a 1 to 2 4/5, (according to Rigaud, as 100 to 270.) The islands form scarcely 1/22 of the continental masses, which are so unequally divided that they consist of three times more land in the northern than in the southern hemisphere; the latter being, therefore, pre-eminently oceanic. From 40° South latitude, to the Antarctic pole, the earth is almost entirely covered with water. The fluid element predominates in like manner between the eastern shores of the old, and the western shores of the new continent, being only interspersed with some insular groups. The learned hydrographer, Fleurieu, has very justly named this vast oceanic basin which, under the tropics, extends over 145° of longitude, the Great Ocean, in contradistinction to all other seas. The southern and western hemispheres (reckoning the latter from the meridian of Teneriffe (*sic.*)) are, therefore, more rich in water than any other region of the whole earth.

These are the main points involved in the consideration of the relative quantity of land and sea, a relation that exercises so important an influence on the distribution of temperature, the variation in atmospheric pressure, the direction of the winds, and the quantity of moisture in the air with which the development of vegetation is so essentially connected. When we consider that nearly three-fourths of the upper surface of our planet is covered with water, we shall be less surprised at the imperfect condition of meteorology before the beginning of the present century; since it is only during the subsequent period that numerous accurate observations on the temperature of the sea at different latitudes, and at different seasons, have been made and numerically compared together (von Humboldt in Maury 1851, 3–4).

The importance of knowledge for knowledge's sake appears in the eyes of von Humboldt as a justifiable purpose and a praiseworthy way of spending one's time. If someone then benefits from this knowledge, that is a bonus. As Maury writes (1851, 4), referring to his *Sailing Directions*:

> It is not for the benefit of navigation alone that seamen are invited to make observations and collect materials for the Wind and current charts; other great interests besides those of commerce have their origin in the ocean or the air; and these interest are to be benefitted by a better knowledge that we now have on the laws which govern the circulation of the atmosphere, and regulate the movements of the aqueous portions of our planet.

The "advance of meteorology in general" is as valuable as advantages in practical navigation, as relations in temperature that can be used to better plan global agriculture (von Humboldt in Maury 1851, 4).

As I have shown here, the importance of Maury's contribution to knowledge of the ocean is manifold. Heralded by the direct applications and non-negligible monetary savings, his publications rapidly gained importance within the scientific community as championing a methodical technique. *Hunt's Merchant's Magazine*, a respected and well-read news chronicle in nineteenth century United States, reviewed his *Sailing Directions* thus:

> We have in this book, with its unpretending and uninviting title, another instance of the great results that spring from small facts and simple ideas. We will not venture to compare the results before us with those that sprang from Newton's apple or Galvani's frog, for from one we have the great law of gravitation, and from the other the magnetic telegraph; but we will venture to assert that since the invention of the mariner's compass, and its great adjunct, the chronometer, no such boon has been given to commerce, and through it to civilization, as this book confers.
>
> The small and apparently unimportant fact that when a commander was appointed, for the first time, to a ship bound from a port in the United States to Rio de Janeiro, he naturally enough asked some other captain, familiar with the route, to point out the best course to steer in order to make the quickest passage, turned over in the philosophic and comprehensive mind of the author, gave birth to the simple idea of collecting and setting forth the experience and knowledge of all.
>
> (Hunt 1854, 531)

As systematic method proved to help predict the best course of sailing, the Enlightenment playing field, once an elitist strata of society, was brought down to the decks of merchant ships and effectively popularized as a new way of knowing where to go. Maury arguably gave literal and philosophical direction to the world's sailing. Through considering Maury's work in relation to the Enlightenment, cartography, and navigation, it thus emerges that the production of knowledge about the ocean-space was negotiated with regard to all of these factors, as well as bearing witness to a more general interrelatedness between different strata of society and economic factors.

4.2.4 Mapping magnetic variation

The final example that I will discuss here with regard to the making of geographical knowledge of the ocean-space during the Enlightenment pertains to magnetism. Both the circumstances under which this phenomenon was mapped and its implications for navigation embody the way that the period's thinkers dealt with the earth as an object of geographical inquiry. Here I consider the discovery and navigational uses of this magnetic field that helps land and sea vehicles locate themselves and steer steady courses. This will first require the introduction of certain notions about geomagnetism so that its uses with regards to navigation might be made clearer.

In its simplest form, the earth's geomagnetic field can be conceived of as an elongated magnet that runs from the Magnetic North Pole to the Magnetic South Pole through the center of the earth.[16] This causes a magnetic field to exist on the surface of the planet, which, thanks to the close proximity of the Magnetic North and South Poles to the Geographic North and South Poles, can be used to obtain geographical directions. Adjusted to include the deviation between the two north poles, geographical direction can be easily deduced with a magnetized needle that can align itself with the field. This is the basic principle of the magnetic compass and, while such simple compasses may not be very precise, they can nonetheless provide accurate readings of compass points that can be used to steer or orient oneself with relative simplicity. Furthermore, the magnetic readings can themselves be adjusted, with the correct information, to produce more detailed readings if necessary.

The history of the magnetic compass is a long and convoluted one which bears numerous elements of legend.[17] Furthermore, the invention of the compass had to be preceded by a series of discoveries, which, when harnessed, would lead to the invention of the needle-and-bowl magnetic device. Holdore and May (1973, 43) list them:

> [First], the discovery that there existed a certain ore, named lodestone, which had the property of attracting to it pieces of ferrous metals; second, the discovery that this power of attraction, known as magnetism, could be transmitted to such pieces of ferrous metal; third, the discovery that there was some curious property by which certain magnetic objects appeared to repel instead of to attract each other; and, finally, the discovery that a piece of magnetized metal of suitable shape would, if freely suspended, point towards the north.

Significantly, this sequence of moments of discovery highlights the processes that led to the eventual invention of the magnetic compass and the ultimate progress that the device led to navigating the world's ocean. However, rather than looking at the history of the device itself, I will focus here on the discovery and mapping of earth's magnetic fields and those uses in navigation and in relation to knowledge.

In particular, other than in its use to orientate the needle of the compass, the earth's magnetic field can be used to locate oneself along a line of equal magnetic deviation. Indeed, because of the shape of the earth's magnetic field, the magnet dips toward the Magnetic North Pole at a degree that varies according to the needle's relative location to the pole. In this manner, magnetic deviation can be used to approximate one's position on a line of equal magnetism, once the appropriate measurements and mappings of the magnetic field have been taken and these results mapped accurately. What this means is that knowing the earth's magnetic field is very useful in terms of navigation at sea; but in order to know these things, data needed to be collected. In 1698, Edmond Halley set sail aboard a pink, HMS *Paramore*, under orders to "improve the knowledge of the Longitude and variations of the Compasse" (Thrower 1981, 268–269).[18] The maps that he produced over the course of his three journeys remain milestones in the histories of science, knowledge, and cartography.

Halley was, according to Samuel Pepys (in Thrower 1981, 15), "the most, if not to be the first Englishman (and possibly any other) that had so much, or (it may be) any competent degree (meeting in them) of the science and practice (both) of navigation." In his diary, Pepys recognizes Halley's "contribution to seamanship, an approach at once theoretical and practical" (Thrower 1981, 15). Indeed, Halley's trips aboard HMS *Paramore* between 1698 and 1701 were extraordinary in several ways. First, Halley's trips are remarkable in that they constitute the first instance of a ship being commissioned for the purpose of a scientific expedition. Indeed, the first of the three journeys was "the first sea journey undertaken for a purely scientific object" (Chapman in Thrower 1981, 69). Indeed, Halley "planned and captained [his] extended voyage to test his own theories *without an idea on his part of commercial advantage*," and in so doing conducted the first ship-based scientific expedition (Thrower 1981, 16, my emphasis). This was 150 years before Maury's work on currents, which, for all that they championed knowledge for knowledge's sake, were, as shown earlier, rapidly praised for their commercial applications and popular because of the direct economies they provided.

Also unique to Halley's first journey is the fact that he was not a military man and yet was in charge of a military embarkation. This caused tensions onboard the ship, and, consequently, the expedition had to be interrupted so that Halley could be given a temporary captain's rank in the Royal Navy and thus acquire authority over his naval crew. This contextualizes the expedition within a larger framework of seafaring that, to a certain extent, permeates through to the present: namely that the sea, its mapping, its control, its authority is military. Similarly, as seen earlier, the discussion between captains of English packets and American fisherman was likely affected by this perception. In the case of Halley, his rank, albeit temporary, was crucial to his being able to lead HMS *Paramore*'s expedition. Today, the military's influence on the mapping of the ocean remains strong, as the United Kingdom Hydrographic Office remains a department of the Ministry of Defence.[19]

Other national hydrographic offices, such as the French Service Hydrographique et Océanographique de la Marine (SHOM), are also today military. It is undeniable that the military/civilian dichotomy underpins a certain way of producing knowledge about the ocean-space and has influenced a number of aspects of knowing the ocean-space.

Despite the initial unrest onboard, though, Halley still produced results after his first trip:

> Mr. Halley shewd the several Variations of the needle he had observed in his voyage, sett out in a Sea chart, as also he shew'd that Brazile was ill placed in the Comon Mapps, and he shewd some Barnakles which he observed to be quick of growth.
>
> (Thrower 1981, 295)

Standing alone, this is already no small feat. Halley's next two journeys would further adjust maps but especially diligently measure magnetic deviation across the ocean and fulfill his instructions to "observe . . . the variations of the Compasse, with all the accuracy that you can" (Thrower 1981, 269).[20]

The method Halley used to determine magnetic variation is simple enough: it is the difference in degrees shown between an azimuth compass and a magnetic compass.[21] However, these measures need to be taken as precisely as possible and, preferably combining at least two readings. By taking as many readings as possible across the ocean, Halley was able to map with some precision magnetic deviation in the Atlantic Ocean and thereafter extrapolate lines of the same magnetic deviation. He then charted those lines on a map of the Atlantic Ocean, which was published upon his return to Britain (Figure 4.2). This map represents the most important piece of work to emerge from Halley's journeys in terms of tangible contributions to the body of knowledge.

The chart has two cartouches[22]; one explains what the map does:

> The curve lines which are drawn over the Seas in this chart, do shew at one View all the places where the Variation of the compass is the same; The numbers to them shows how many Degrees the needle declines either eastwards or Westwards from the true north; and the Double line passing near Bermudas and the Cape de Virde Isles is that where the needle stands true, without Variation.

The other cartouche dedicates the chart to King William III and titles the map in Latin. While no date of publication appears on the map, it is generally accepted that it first appeared in 1701, and also incorporated the data from Halley's second trip, which returned to England on 18 September 1700.[23] Though it is one of the first maps of the eighteenth century, it embodies several characteristics that

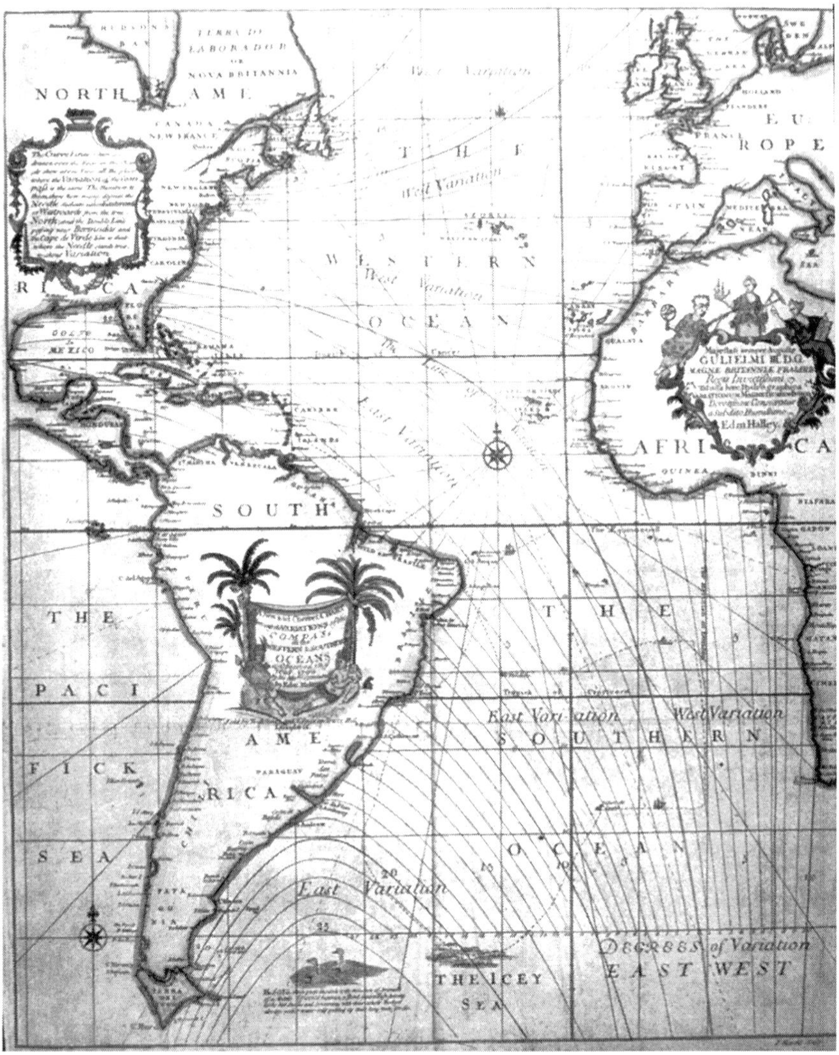

Figure 4.2 Halley's chart showing the variations of the compass in the Western and Southern Oceans, c. 1701. The original chart is a hand-colored copper engraving measuring 58.5 × 49 centimeters.

Source: Royal Geographical Society Picture Library.

cartography would take over the next hundred years in terms of changing the way that data were presented.

The eighteenth century witnessed a variety of changes in the field of visual representation (Friendly 2009). The century was a period during which states felt the need to measure their population, accumulate data about their assets, and represent these results in usable manners that developed from theories of probability. It was the period that saw the "invention of 'population' as a scientific category (. . .), an entity which could be enumerated and analyzed in the new language of statistics" (Heffernan 1999, 133). This new way of thinking had an impact on visual representation. As Friendly (2009, 1) writes:

> This birth of statistical thinking was also accompanied by a rise in visual thinking: diagrams were used to illustrate mathematical proofs and functions; nomograms were developed to aid calculations; various graphic forms were invented to make the properties of empirical numbers—their trends, tendencies, and distributions—more easily communicated, or accessible to visual inspection. As well, the close relation of the numbers of the state (the origin of the word "statistics") and its geography gave rise to the visual representation of such data on maps, now called "thematic cartography."

He continues:

> The 18th century witnessed, and participated in, the initial germination of the seeds of visualization which had been planted earlier. Map-makers began to try to show more than just geographical position on a map. As a result, (*sic*), new graphic forms (isolines and contours) were invented, and thematic mapping of physical quantities took root. Towards the end of this century, we see the first attempts at the thematic mapping of geologic, economic, and medical data.
>
> (Friendly 2009, 10)

The idea was that "data could 'speak to the eyes'" (Friendly 2009, 10). Indeed, the eighteenth century changed the way that data was conceived of and altered the manner in which it was made sense of visually. Examples of this were shown earlier in the discussion of the markings next to Isle Grande and the cohabitation of ships' tracks that were measured through different methods.

In this context, Halley's 1701 map is a landmark in many ways. Indeed, while showing—and correcting—geographical knowledge (the coast of Brazil, in particular), the core of the map is not primarily geographical. By showing averaged lines of the same magnetic variation, Halley impresses upon the globe a set of measurable and mappable phenomena that is beyond the realm of geography as such. For the first time, the map depicts data that is not simply about geographical locations and relative positions: this is where Halley originally made use of what has become known as isogonic, or Halleyian, lines. Here we see "lines of equal magnetic declination for the world, possibly the first contour map of a data-based variable" (Friendly 2009, 10). Soon, similar lines would become useful in charting other physical geographical features such as depth (isobath), the magnetic dip of the needle (isocline), atmospheric pressure (isobar), and surface temperature (isotherm).[24] Generically, these are called isopleths or isograms, and cartographically, their importance lies in that they give two-dimensional maps a way of portraying data that operates in three-dimensions or is invisible. Furthermore, Halley's chosen manner of representing knowledge falls within a wider Enlightenment project of "structuring space and classifying geographical information" (Edney 1999, 176). Halley was, like von Humboldt:

> looking for commensurability between the richness of nature and a language capable of representing and ordering that richness. He was looking for a language at once highly descriptive but also analytical: capable of moving beyond geographic space to assume the spaces of scientific theory; capable of comparing like things in very different places; and the one that could reveal new linkages.
>
> (Godlewska 1999, 267)

This language drew itself through these new graphic lines. As such, Halley's map epitomizes scientific method, but also marks a milestone in the development of how data is used and presented. Thus it bears witness to the twin Enlightenment project of mathematicizing phenomena and making sense of the earth as a geographical entity.

4.3 Conclusion

This chapter identified certain processes through which the Atlantic Ocean was turned into facts, how data about it was made, and how measurements became knowledge. With specific attention to the Atlantic Ocean, I highlighted how the ocean-space became perceived as "an object of Enlightenment enquiry," as part of the Enlightenment project that sought to know the earth scientifically (Withers 2006b, 72). The respective cases of nonexistent islands, coordinates, ocean currents and magnetism in the Atlantic Ocean were analyzed against the background of Enlightenment geography where the ocean-space emerges as a scientific space of circulation. By interpreting the ocean-space geographically, I have shown how it was, during the Enlightenment, a space of entwinement of peoples, flows, and networks, all of which defined it within these circulations. Through their work, Franklin, Maury, and Halley not only advanced knowledge but also participated in changing the nature of knowledge itself, making it provable, strict, and scientific. In fact, the charts examined can be considered foremost as "records of evolving geographical knowledge of the world rather than with their characteristics and use as navigational tools" (Akerman 2006b, 9). The facts they represented were dependent on timeless principles rather than ephemeral cultural trends, as the charts themselves are representative of knowledge, which is verifiable rather than apocryphal and situated within a broader discourse of scientific inquiry. Even in the case of mapping the nonexistent islands, it became evident that cartographers tested and sought to verify their facts. Indeed, knowledge was progressive here, as "an episode in science is progressive when at the end of the episode there is more knowledge than at the beginning" (Bird 2007, 64). These advances were translated cartographically into more accurate maps that not only showed the place of coasts and islands but also favorable routes to travel the seas. Moreover, the techniques of cartography and the mathematical lens through which the earth became charted purveyed a further sense of scientific improvement. Navigation itself became more grounded in science as currents and winds were, with the help of Maury, capitalized upon by sea captains and merchant navies. This had economic ramifications that tied into the interests of the European empires and thus encouraged more such work. By thinking about the Enlightenment in terms of geography and with regard to the ocean-space, I also located this discussion within a wider set of debates that questions spaces of knowledge: here, indeed, knowledge has locales that shape it. The question "where was the Enlightenment?" has a plethora of answers, and this chapter demonstrates how the Enlightenment was located in specific networks of trust (Withers 2007, 11). At the end of the period I have examined, the ocean-space, and in this case the Atlantic ocean-space, was known through science, and maps translated this new knowledge into charts that would stand the test of time because they were based on sources that were scientific, rather than hearsay. The following chapter will now turn to consider how knowledge about the ocean-space acquired a globally scientific scale by considering the deep sea and specific scientific enterprises that explored it.

Notes

1 This coordinates are in relation to the Greenwich meridian. Loyasa's map would have likely used the Cádiz or Lisbon meridian, 6 degrees 8 minutes and 9 degrees 11 minutes West of Greenwich respectively.
2 UKHO Survey B. 424 5. G: *Laurie and Whittle's Chart of South America and the Southern Ocean; Including the Western Coast of Africa, from Cape Verd to the Cape of Good Hope*, 1805.
3 UKHO Survey C. 138: *The Track of His Majesty's Sloop Ship* Julia, *Jenkin Jones Esq. Commander, in search of the Island of St. Matthew*, 1817.
4 UKHO OCB. 357 A. 1: *A Chart of the Ethiopic or Southern Ocean, and part of the Pacific Ocean; from the Parallel of 3 degrees North to 56° 20' South Latitude and from 20° East to 90' West Longitude drawn from the latest observations of the Spanish, Portuguese and Dutch Astronomers, shewing the track of the Warley, East Indiaman, outward and homeward in the years 1805 & 6*, 1808.
5 UKHO Survey Z1 (shelf Hf): *Carta Esferica del Oceano Meridional desde el Equador haste 60 grados de latitud y desde el Cabo de Hornos hasta el Canal de Mozambique construida de orden del rey en la dirreccion de trabajos hidrographicos y presentada á S. M. por mano del excmo. Señor Don Antonio Cornel, Descretario de Estado, y del Despacho Universal de Guerra, encargado del de Marina, y de la Direccion general de la Armada, año de 1800.*
6 Sailing lore says that when measuring longitude with a sextant, one can start feeling confident only after one's first thousand calculations have been wrong.
7 UKHO Survey A. 634 Ag. 4: *Plano del Puerto y Ciudad de La Havana levantado por D. José del Rio, Capitan de Fragate de la R^l. Armada*, ano de 1798.
8 UKHO Survey E. 88. Ag. 3: *Plan of the Harbour and City of the Havana, Surveyed by Don Joseph del Rio, Captain in the Spanish Navy, 1798, republished on the scale of the original plan, by W. Faden, Geographer to His Majesty and to His Royal Highness the Prince of Wales*, 1805.
9 Dead reckoning is a method by which the navigator estimates a ship's position by using a series of calculated and estimated data. In the seventeenth and eighteenth centuries,

this was limited to averaging the speed of the boat and the ship's heading as measured with a compass. As more information was gathered about the ocean, such as currents, the mechanics of fluids, naval technology, and the earth and how it affects instruments (such as magnetic deviation), the technique of dead reckoning was fine-tuned and is today very accurate if performed rigorously. However, with the limited knowledge of the time, dead reckoning was very imprecise.

10 Latitude navigation is based on the idea of keeping a ship's course parallel to a chosen line of latitude. When a ship made it to the desired line, it sought to stay on it, doing so by keeping certain stars at the same height in the sky. This method was reliable, if limiting, but was the one used by Christopher Columbus to attempt to reach the East Indies in 1492. If the Americas had not existed, he would have probably succeeded, possibly landing around Shanghai.

11 UKHO Survey B. 621/1–4 (shelf Cp): *To the right Hororable Charles Lord Barham, First Lord of the Admiralty, First Commissioner for Revising the civil offices of the Navy, Admiral of the White Squadron of His Majesty's fleet &c &c &c. This Chart of the Atlantic Ocean is most respectfully dedicated, by his Lordship's Most Humble Servant, A. Arrowsmith, Hydrographer to His Majesty*, 1805 with additions in 1812.

12 UKHO Chart B. 421 3. D: *A New General Chart of the Atlantic or Western Ocean and Adjacent Seas, including the coasts of Europe and Africa from 60 Degrees North Latitude to the Equator and also the opposite coast of America; Drawn and Regulated by the most accurate astronomic observations, and the journals of the most experienced navigators in which are also particularly distinguished the determinations of the longitude as given by the marine clocks of Mr. Ferdinand Berthoud in the voyage performed by order of the government of France in 1768 and 1769 in the ship Isis to the Azores, Madeira, the Canaries, Cape Verd Islands, St. Domingo, the Great Bank of Newfoundland, &ct by Mr. de Fleurieu, on officer of the French Navy and Member of the Royal Marine Academy, &ct*, 1803.

13 The letter, known as "Franklin's Sundry Maritime Observations," contains thoughts on the action of the wind, new sails that minimize wind resistance, anchors, accidents at sea, and sea-worthy soup bowls. It has been written that "if this one letter had been all that survived of Franklin's correspondence, he would have been remembered as a scientist of note and an individual with a remarkable span of knowledge on many subjects" (NOAA 2009).

14 In this passage, where "gulf" is spelt "gulph" I have left it as such.

15 While this system of hoping that a specific bottle launched somewhere in the ocean turns up somewhere else at a later date is very primitive, it remains one of the most reliable and accurate ways to trace surface currents. More recently, the accidental loss of bath toys and shoes has also helped map ocean currents.

16 This is an approximation: In reality, the Magnetic North and South Poles are not antipodal: the Magnetic South Pole is farther away from the Geographic South Pole than the Magnetic North Pole is from the Geographic North Pole.

17 Poet Nicander of Colophon tells of the story of a cowherd who became immobilized when the iron nails in his sandals adhered to the rock (Holdore and May 1973, 43).

18 A pink is a type of flat-bottomed ship with a narrow stern.

19 The military nature of the UKHO had minor implications for the research undertaken in their library for the thesis on which this book is based. First, not being a British citizen, the author had a longer waiting period before she would be granted access. Then, the site itself is a heavily guarded space, and passport details and several forms of identification were required to access the site. Once on site, the author had to be escorted everywhere on site until she upgraded her access badge by providing some ten pages worth of personal information. Certainly, the documents examined in this context did not warrant such security, but their affiliation to the army and the fact that they are on a military site highlighted the fact that they were once the pinnacle of military missions. In contrast, the United States equivalent, the National Oceanic and Atmospheric

Administration (NOAA), is a federal agency under the purview of the Department of Commerce. Security checks to consult their archives were not as stringent: all that was needed to consult their library was identification and my laptop's serial number.

20 While Halley indeed completed three journeys, I will only discuss the findings of the first two here, as the third one was concerned with research in the Channel: re-mapping the coasts of England and France, and tidal measurements.

21 An azimuthal method of determining direction uses the night sky. This can be as simple as using the North Star.

22 *A New and Correct Chart Shewing the Variations of the Compass in the Western & Southern Oceans as Observed in ye year 1700 by his Maties Command by Edm. Halley.*

23 Furthermore, the dedication to King William III, who died on 8 March 1702, supports this theory (Thrower 1981, 56).

24 Also, isotach: line showing equal wind velocity; isopach: line showing equal rock thickness; isoseismal: line showing equal strength of earthquake; and isopycnal line: line connecting points where the water has the same density.

5 In-depth knowledge

The deep ocean-space

In Chapter 3, I discussed the mental discovery of the Atlantic Ocean as the cognitive process through which the Atlantic ocean-space became known in cultural terms and accepted according to a new, science-based paradigm of knowledge. This mental discovery of the Atlantic Ocean was the first step toward knowing the global ocean-space as a geographical space. Chapter 4 examined the lateral borders to the ocean-space, with examples from the Atlantic Ocean, and described how it was mathematicized and the discovery and negotiation of specific facts about it. In particular, I examined how work on ocean currents and the earth's magnetism were instrumental in bettering navigational passages on the ocean's surface. By the end of the Enlightenment, the Atlantic Ocean was known as a geographical space with defined limits and having measured phenomena. This chapter will study how the ocean's depths and the seabed became understood both culturally and scientifically. Indeed, both the ocean's depth and the nature of the sea floor had to be discovered before they could be scientifically explored. First, I discuss the idea of the ocean floor as a three-dimensional space. Certainly, even though little was known about the seabed, acknowledging its existence signified that the ocean-space was finite. Second, I will examine efforts to mathematicize this space by sounding its depths; these enterprises will be located within the broader context of "big sciences" and "the geography of knowledge" (Harris 1998, 269; see also Burstyn 2001). Through considering questions of scale with regard to collecting, organizing, and classifying of knowledge about the seabed, I will highlight certain issues regarding the deep ocean-space. Here, as Withers (2007, 183) puts it, "geographical knowledge was central to thinking about the human place in nature." To conclude, I will set my discussion against discourses of scale, returning to debates raised earlier about the locales of science. In the end, this chapter will have used various tropes to highlight the gradual process that led to thinking about the ocean-space as a geographical whole.

5.1 The idea of the deep ocean

Interest in the ocean's depths is, compared to the ocean's surface and navigable routes, relatively recent: Rozwadowski (2005, 7 and 4) notes that "before the last quarter of the eighteenth century, understanding the ocean's depths derived mostly from the imagination," adding that "before the nineteenth century, [the] deep sea made hardly any impression on most people, even citizens of maritime nations." However, this is hardly surprising because, from a practical point of view, there was very little need to know the ocean's depth: once safe, navigable sailing routes were established, there was no need to stray beyond these. Yet, by the late-nineteenth century, for a variety of reasons that will be discussed in a moment, expeditions were being set up to sound the ocean deep, overcoming both practical and technical barriers in doing so. First, however, I will discuss the concept of the deep ocean as a space that needed to be understood culturally. I argue that the deep ocean-space was a space that had to be wholly imagined. Even the first soundings did not help create an image of what the deep ocean might look like and, together, the deep ocean and the sea floor constituted a complete unknown, and the first step toward making them known was mental. Before technology could give dimensions to the deep sea, minds needed to understand it as a new space.

For the purposes of navigation, extensive and precise knowledge of the ocean's depth was not essential. Located soundings were useful to avoid running aground in specific areas, and regular soundings could be used to detect a continental shelf, "which gave seamen advanced notice of the proximity of land after long ocean voyages" (Laughton 2001, 92). However, as ships often navigated along known routes, more knowledge about the depths beyond navigable routes was superfluous. Indeed, as Henry David Thoreau (in Rozwadowski 2005, 66) pragmatically put it,

> of what use is a bottom if it is out of sight, if it is two or three miles from the surface, and you are to be drowned so long before you get to it, though it were made of the same stuff with your native soil.

In essence, "the depth of the ocean was of only academic interest—to be argued but never measured" (Laughton 2001, 92). Until the mid-nineteenth century, there was therefore a strong sense that sounding and exploring the ocean's depths was an expensive folly rather than a practical enterprise. Therefore, ideas about what the sea floor might look like were usually theoretical.

At the time, one popular theory that sought to conceptualize the ocean's depth was comparative: the ocean was probably deep in the same way that mountains are high, and the sea floor would be a kind of negative of land. Even though this approach had no empirical grounding, it was popular. It explained shallow waters, steep falls, and depressions that had been remotely observed with rope and wire, and tentative profile maps such as Maury's 1855 "Vertical Section—North Atlantic" were reminiscent of terrestrial profile maps (Figure 5.1). From this perspective, the deep sea was comparable to alpine mountains: sublime and enthralling, if unattainable and dangerous. Yet, while this simple comparative approach provided a model to visualize the deep sea and the seabed, it also designated it as mysterious and unattainable, as mountain summits had once been.

The idea of an unattainable deep sea was corroborated by a popular belief that the deep sea was in fact physically unreachable. One element of this theory can be traced to a refutation of Henry Fox Talbot's 1833 suggestion of using "an exploding shell to propagate sound from the floor of the sea to the surface" and measuring

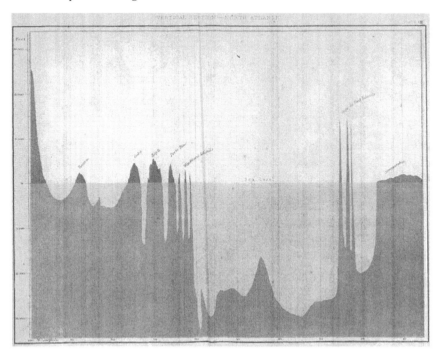

Figure 5.1 "Vertical Section—North Atlantic" (Maury 1855 [2003], Plate VIII).

the sound wave caused by the explosion to deduce measurements (Deacon 1997, 285).[1] The criticism postulated that Talbot's idea would be ineffectual because, at great depths, the density of water increased with pressure. Therefore,

> Mr. Talbot's shell would *float* long before it could reach the bottom, as it is now pretty well ascertained that at a certain depth the sea is specifically heavier than any body of which we are acquainted with, consequently a cast-iron shell could not penetrate it.
>
> (In Deacon 1997, 285)

Offshoots of the idea of impenetrable depths led to peculiar beliefs, which propounded that objects that fell in the sea hover at a certain depth forever. (Of course, it is true that pressure increases as depth increases, but not to the degree suggested here.) Thomson, recalling this then-obsolete belief while aboard HMS *Challenger* as chief scientist, reminisced that:

> There was a curious notion, in which I well remember sharing when a boy, that, in going down, the sea-water became gradually under the pressure heavier and heavier, and that all the loose things in the sea floated at different levels, according to their specific weight: skeletons of men, anchors and shot and cannon, and last of all broad gold pieces wrecked in the loss of many a

galleon on the Spanish Main . . . beneath which there lay all the depth of clear still water, which was heavier than molten gold.

(Thomson in Deacon 1997, 285)

This idea, though unfounded and unverifiable, seeped into cultural consciousness, shaping a concept of the deep as physically unattainable and the space between as weirdly populated by objects long forgotten. For instance, Jules Verne talks about these floating sunken ships in *Twenty Thousand Leagues Under the Sea* (Verne 1990). However, rather than shaping or hindering the advancement of science, these ideas were mainly limited to the sociocultural sphere, affecting the popular rather than scientific understanding. However, it is interesting to note the tension between science and culture, and how knowledge passes between them: the idea of the deep sea as an unattainable space seeped through social spheres and into cultural consciousness despite being refuted in scientific circles.

Certainly, Verne's description of the deep in his novel bears witness to this, and it is undeniable that literature plays a significant role in shaping the idea of the sea in the cultural realm. This is comparable to what happened with Franklin's and Maury's respective works on currents, which similarly infiltrated the lower social classes through a popular merchants' magazine. However, unlike the idea of impenetrable pressure in the deep sea, Franklin and Maury were correct about water movements. What I will argue here is how the scientific idea of mapping the deep "served (. . .) as an intellectual base line" in the nineteenth century, parallel-ing the formation of thought about the deep sea to that of the space of coastlines during the Enlightenment (Carter 1999a, 132). Here the mapping of coastlines provides a useful analogy for comparing attempts at mapping liminal spaces that act as a border between the known and the unknown.

The coast, Carter (1999a, 125) writes, is "an a priori of geographical discourse, a meta-geographical proposition enabling one to say something valuable about the geographical phenomena found along it." It thus emerges as a somewhat inde-terminate, unreachable space that was representative both of itself and of geo-graphical knowledge it beholds.[2] Similarly, while it was not directly visible like the coast, the seabed was assumed in geographical thought as a phenomenon that allowed something to be said about the ocean-space. In the same way that it was accepted that mountains had summits, logic stated that the sea must have a bot-tom. Presumably too, while it had not been reached yet, the idea of a sea floor must have given more solace than the idea of a lack of seabed. Thus the sup-posed existence of the sea floor epistemologically allowed for the ocean-space to be conceptualized as a finite whole that could be measured and known. As an abstract yet defined concept, the sea floor provided a physical as well as philo-sophical foundation to think about the ocean-space as an entity. Indeed, the sea floor was, like the coast,

a generalization, an abstraction: but as the medium of connecting isolated objects to one another, it was a condition of knowledge, an analogue of the associative reasoning essential to the orderly progress of reason (. . .).

(Carter 1999a, 125)

As such, the conceptual line of the ocean floor was both a mental hypothesis and an ontological tool to make sense of the vast unknown of the ocean-space.[3] Unknowns or imprecisions about the line that defined the deep provided an epistemological motivation to reason, measure, and map the ocean-space. As Maury (2003, 282) writes:

> Astronomers had measured the volumes and weighed the masses of the most distant planets, and increased thereby the stock of human knowledge. Was it creditable to the age that the depths of the sea should remain in the category of an unsolved problem? Its "ooze and bottom" was a sealed volume, rich with ancient and eloquent legends, and suggestive of many an instructive lesson that might be useful and profitable to man. The seal which covered it was of rolling waves many thousand feet in thickness. Could it be broken?

The discovery of the ocean's depth therefore represented an enterprise bigger than itself.

Discovering and being able to map the sea floor or the coastline represented more than the specific line that it drew. As Carter (1999a, 126) notes, drawing these lines stood for the "command of getting from one place to another without any gaps." For Carter, mapping the coastline represents making the unknown less so by limiting it with lines and locating it on a grid. On a chart, the coast's line can thus demarcate the known from the unknown. In terms of deep-sea sounding, collecting data about the unreachable sea floor, however imprecise and unverifiable, also constituted progress: if not factual, at least technical and philosophical. In this sense, both the coastline and the isobath of the deep sea stand in for the limits of knowledge as well as for knowledge itself. Isobaths, like coastlines before them, epitomized the paradox of the Enlightenment method. Indeed, "Enlightenment geographical discourse aspired to certainty (. . .) although its own procedures were irreducibly inductive" (Carter 1999b, 295). In other words, drawing them on charts represented one aspect of knowing the ocean-space, while highlighting the yet unknown aspect of its specific characteristics. As Carter (1999a, 128–129) put it concerning the coastline,

> A science that claimed to be well founded had *ipso facto* to leave no gaps in its reasoning, but if it were to make any progress it had necessarily to begin somewhere and end somewhere else—without a little *saltus* of imagination, the gift of intellectual wit, or what Vico called *ingenium*, to see relationships and grasp their significance, no progress was possible. But for the gap there was nothing to know, but the scope of knowledge was to eliminate it. And the instrument of elimination is the continuous line.

In effect, the practice of surveying and mapping a coast or a sea floor is as much of a physical process as an ontological one, where known and unknown compete with each other for cartographic and philosophical space. Concretely, this acknowledges that soundings could not give any indication of the properties of the seabed on either side of the point where the sounding weight touched the bottom. When cartographers drew lines and profiles, the links between the sounding stations were little more than glorified guesswork. Thus the seabed had to be conceptualized in a linear dimension whose profile was, as the coastline had been before, "as much an intuitive aspect of chart-making as a representation of what had been seen" (Carter 1999a, 125–126). This prospective conception of the sea floor was an imprecise and unverifiable one, but it did provide a basis for future soundings. As when coasts were misplaced and then relocated when more data or more precise instruments were available, when unsounded depths were measured, they were corrected on the map.

Thus mapping the coast or drawing the seabed's profile are two mapping enterprises defined by their transitory character, which motivated the search for final, viable data. Both the coast and the ocean floor "kept alive the dialectic between the seen and the unseen whose rationalization was critical to the formulation of Enlightenment epistemology" (Carter 1999a, 136). The epistemological processes at play to hypothesize the shape of the sea floor encapsulated the spirit of the Enlightenment, which optimistically sought to advance knowledge by all means, despite some inherent contradictions in methodologies. Certainly, the discovery of something that cannot be seen necessitates the making of a knowledge gap before that gap can be filled, and the enterprise of discovering and mapping the seabed illustrates this: the need to speculate about the sea floor was necessary in order to be able to measure it. As with the mental discovery of the Atlantic Ocean discussed in Chapter 3, the discovery of the sea floor was a physical as well as a cognitive one. Also, the problems linked to making knowledge about faraway spaces are evident here. In particular, the distance between both the seabed and the ship conducting the soundings, and the seabed and the centers of calculation on land were unique and posed a myriad of problems. This will be discussed in relation to the story of *Bathybius haeckelii*. In all events, geographical imaginations and physical realities diverged, and the nature of the space studied presented practical difficulties that needed to be overcome. What will be discussed now, then, is the technical and scientific process that accompanied the methodological and philosophical matters at play, and the physical and technical aspects of this discovery.

5.2 Sounding the ocean's depths

The physical process of surveying the sea's depth is not comparable to that of surveying land or coastlines. While the objective is, in both cases, to draw a topographic profile of a given geographical area, the physical differences between the settings require distinct tools and methods. The physical nature of sounding the deep sea remains an enterprise rife with physical and technical difficulties. Especially, the impossibility of exploring the seabed in person beyond certain depths means that knowledge about the ocean's depth and floor is remote: samples are collected with miles of rope or robotic devices and depths were measured first with rope and later with echo and satellite technology. Thus knowledge of the deep is based on once removed scientific measurement, and any picture of the deep must be collated from individual samples. Indeed, when dealing with the deep sea, humans will have to deal with the fact that "access below a few feet underwater will always rely on harnessing technologies to overcome [mankind's] limitations as terrestrial, air-breathing animals" (Earle 2005, ix). The crucial role of technology, therefore, cannot be underplayed in the efforts to delve into the deep and reveal facts about the ocean's depths. This technological aspect of sounding the deep sea is what I will consider first here.

Technology, as Summerhayes (2001, 89) writes,

> has been the key to mankind's understanding of the ocean floor since earliest times . . . essentially, primitive technologies, such as the lead-line sounding used in HMS *Challenger*, have in the twentieth century given way to widespread use, first of single-beam echo-sounding, then of multi-beam echo-sounding, and now of interpretations based on satellite altimetry.

This brief summary of deep-sea sounding technologies illustrates the creativity of scientists to obtain a measurement upon which they can build. Indeed, as Deacon (1997, 3) reminds us, "[accurate] measurement is a prerequisite of oceanography." As discussed in Chapter 4, measuring specific characteristics and fixing the limits of the Atlantic Ocean was at the core of its discovery. Yet the limitations of the technology often exacerbate the futility of the enterprise rather than clarify the lay of the sea. Indeed, until the advent of echo sounding, the only working method available to sound the oceanic deep was with rope or wire, a technique that was imprecise, temperamental, and very localized. The method was this: a weighted rope was measured out and marked at intervals of, typically, one hundred fathoms.[4] It would then be unrolled from its storage reel while the crew then measured the time lag between each marking. After timings were adjusted for increased friction, it was concluded that the weight had reached the seabed when the time lapse between each marker became disproportionately longer. However, there was no way of verifying this: the rope could have been caught in a current, or, if it did reach the seabed, it might not rise to the surface vertically, and there was no way of ascertaining this. Furthermore, as mentioned earlier, whatever measurement was obtained was very localized: it was only an indication of depth at a specific spot and could not be used to reliably deduce the lay of the seabed around this place. Certainly, even if a sounding was perfectly accurate, it remained inherently limited. Highlighting these limitations, George Wallich (in Rozwadowski 2005, 69) wrote after observing soundings from aboard the HMS *Bulldog* in 1860:

> Suppose a person were to walk across the Welsh Mountains from one end to the other in a straight line or nearly so, and were blindfolded, to dig into the ground beneath his feet at certain intervals a cylinder like that of Brooke's machine. Could he by any possibility in this manner obtain an average sample of the soils or surface structure he would pass across? Most certainly not.

As Rozwadowski (2005, 69) writes: "Wallich's point appeared as ludicrous for geology in the mid-nineteenth century as it does today." However, at the time, rope sounding was the only solution: while the technique was fully understood as imperfect, the gaps that it produced were as useful as the holes that it filled. In this respect, the use of rope to sound the ocean and attempt to obtain a general picture of the seabed exemplifies the optimism of the scientific method and embodies the drive to make facts and measure the epoch. Nonetheless, with the advent of deep-sea telegraphy, demand for more precise soundings and the need to know the nature of the sea floor, and as technology improved, more accurate measurements became possible.

In particular, being able to sound accurately and reliably required the capacity for a sounding vessel to remain in the same geographical position for extended periods of time. In order for the sounding to be precise, and for it to be correctly mapped, avoiding drift was essential. When ships were solely wind powered, this was impossible, as they could not stabilize against drift. However, as ships were fitted with steam engines, countering drift became possible. For instance, HMS *Challenger* was equipped with a twelve-hundred-horsepower supplementary engine for the purposes of dredging, sounding, harbor maneuvers, and emergencies (Linklater 1974, 15). For all other activities, HMS *Challenger* remained a sailing vessel, and her screw was lifted out of water while she was under sail (Linklater 1974, 15). Therefore, by enabling a ship to remain stationary in relation to the sea floor, the steam engine was unequivocally a key actor in fathoming the ocean.

Moreover, steam technology also facilitated the handling of the many lengths of rope required for the sounding process (Rice 2001, 35). For example, HMS *Challenger* alone had

> a staggering total of 220,000 fathoms (more than 400 km), ranging from relatively thin (but unspecified) current drogue line to 3 inch (*c.* 75 mm) circumference dredging rope; during the voyage she "expended", that is lost or discarded, no less than 125,000 fathoms.
>
> (Rice 2001, 31)

Therefore, the advantages of supplementary power essential to operate the wheels on which the rope was stored are obvious.

As noted earlier, however, obtaining a measurement of the ocean's depth was not the only objective of these sounding operations: they also sought to determine the nature of the seabed. There were two reasons for this: one was practical and the other philosophical. The first was linked to the spread of telegraphy, which involved laying cables on the seabed. Cabling companies needed to know how deep the sea floor was in order to measure out the correct amounts of cable and whether or not the seabed would damage the cables themselves if they were too rough and rocky. The second reason was more philosophical in that it was then widely believed "that the ocean's depths would yield secrets about the origin and distribution of life" (Rozwadowski 1996, 411). Much excitement and curiosity

often accompanied sounding, from both scientific and nonscientific crew. As Rozwadowski (1996, 417) writes:

> Initially, dredging, trawling, and sounding inspired intense curiosity about
> what lay beneath the waves. Early in the *Challenger* cruise, each dredge
> haul attracted a crowd of "every man and boy in the ship who could possi-
> bly slip away," each waiting breathlessly for a glimpse of the secrets of the
> deep. Instead of merfolk or monsters, however, sand and mud, and a soon-
> monotonous assemblage of animals appeared in nets and sounding devices.

When the technology was developed to be able to collect and bring up deep-sea sediment, an aura lingered on deck that these would somehow give answers about the
nature of the earth, its origins, and life on it. The discrepancy between the expectations attached to the nature of the findings and the imprecise technology embodies the
optimistic spirit of the Enlightenment that qualified as science during the nineteenth
century. Indeed, while the method had many shortcomings, the period during which
rope or wire sounding was the only option bears witness to the desire to know a
space "where the known and the unknown resembled each other, and might so easily
collapse into the blindness of the same" (Carter 1999a, 147). The desire to measure
and quantify resulted from a new driving force that equated numbers to knowledge.
Thus, however imperfect numbers may be, what mattered for those who were dredging and fathoming was the act of science they were performing. To the scientists
aboard, every time a sounding was successful or ooze was brought up from the deep
sea, it was knowledge that was being advanced. They were showing that "science is
not an abstract body of knowledge, but represents the human activity of observing
and interpreting independently existing complex natural phenomena," even though
they had limited tools to do so (Deacon, Rice, and Summerhayes 2001b, 3). Thus
sounding the deep sea, from practical, technical, and epistemological perspectives
represents a complex enterprise that, in a nineteenth-century "big science" context,
was a "supreme outward expression of [a] culture's aspirations" to know the ocean's
depths (Burstyn 2001, 49; Weinberg in Burstyn 2001, 49). Yet the study of the
ocean's depths, not only sought to yield results in other fields and solve biological
enigmas, it also became entwined in a global geological project. This is a key development in the production of knowledge about the ocean-space, which is evidenced
through geological and biological examples that locate the ocean-space within a
wider, global perspective: the discovery of the Mid-Atlantic Ridge and the discovery
and demise of the creature *Bathybius haeckelii* will respectively illustrate how concepts of the geology of the earth as a whole were evolved and the consequences of a
keen desire to decipher the mystery of the origins of life.

5.3 The bigger picture

5.3.1 The mid-Atlantic ridge and the shape of oceans

During the nineteenth century, the sounding of the oceanic depths became part
of a wider, global project of making sense of the earth's physical geography as

a whole. As certain underwater features were discovered and charted, their discoveries sparked debate regarding their origins, how they compared with other
similar features in other places, and their implications for the earth's geology.
Specifically, the discovery of the Mid-Atlantic Ridge, which was first mapped in
1855 by Maury, provoked a discussion of global, geological nature.

Based on a limited number of soundings in the North Atlantic, Maury deduced
and mapped, in 1855, the existence of the Mid-Atlantic Ridge. It appears on his
map "Basin of the North Atlantic Ocean" (Figure 5.2). He wrote in *The Physical
Geography of the Sea:*

> The basin of the Atlantic Ocean, according to the deep-sea soundings made
> by American and English navies, is shown on Plate VII [Here Figure 5.2].
> This plate refers chiefly to that part of the Atlantic which is included within
> our hemisphere. In its entire length, the basin of this sea is a long trough,
> separating the Old World from the New, and extending probably from pole
> to pole.
>
> (Maury 2003, 289)

However, while Maury was correct in his speculation on the existence and the
extent of the ridge, neither he nor his contemporaries could grasp the geological and global significance of the find. In fact, nineteenth-century discoveries of
the shape of the ocean were instrumental in forming global geological thinking

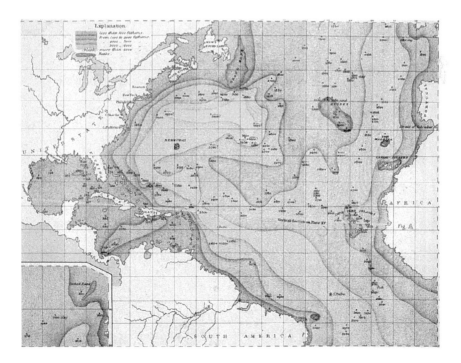

Figure 5.2 "Basin of the North Atlantic Ocean" (Maury 1855 [2003], Plate VII).

and would eventually lead to theories of plate tectonics and continental drift. The Mid-Atlantic Ridge will thus be located within a more general fathoming of the deep ocean and within the wider context of making sense of the earth as a single, geological entity.

Prior to major technological advances of the twentieth-century bathymetric instruments, ocean scientists could do little more than speculate when it came to considering the physical relationship between land and sea in terms of geology, as they lacked the technology to do more. Thus theories were plentiful, but formal knowledge scarce. Yet the first description of the continental shelf and upper continental slope dates from 1706 when Count Luigi Marsigli carried out a series of soundings perpendicular to the southern coast of France (Laughton 2001, 92). He had observed that,

> from the beach, the sea floor fell away for a few fathoms and then levelled out somewhat. It gradually deepened to about 60 or 70 fathoms, at varying distances out to sea, and then fell away sharply to a depth greater than 150 fathoms.
>
> (Deacon 1997, 176)

Marsigli was unaware of the exact nature of what he was describing, and these (almost universal) formations would in fact not be defined scientifically for another 250 years by Bruce Heezen (Laughton 2001, 92). However, though Marsigli's soundings were landmarks in geology and he "recognized the importance of the sea floor in understanding continental geology, [he] surmised that its geology was the same as that of the continents and that the depths compensated for the mountains" (Laughton 2001, 93). However, this theory could not easily be verified and was not pursued. Yet, as seen earlier, the method of comparing underwater features to visible, land-based ones was dominant in early thinking about ocean geology and would prevail until the late-nineteenth century.

Certainly, in the early nineteenth century, Charles Darwin, in his journal from his voyage aboard HMS *Beagle* discussed the correlation between land and the sea floor when speculating about the formation of atolls. He lengthily considered the relationship between atoll island formations, surrounding reefs and the sea floor after observations in the Indian and Pacific Oceans. He writes:

> It is improbable in the highest degree that broad, lofty, isolated, steep-sided banks of sediment, arranged in groups and lines hundreds of leagues in length, could have been deposited in the central and profoundest parts of the Pacific and Indian oceans, at an immense distance from any continent, and where the water is perfectly limpid. It is equally improbable that the elevatory forces should have uplifted throughout the above vast areas, innumerable great rocky banks within 20 to 30 fathoms, or 120 to 180 feet, of the surface of the sea, and not one single point above that level, for where on the whole face of the globe can we find a single chain of mountains, even a few hundred

miles in length, with their many summits rising within a few feet of a given level, and not one pinnacle above it?

(Darwin 1958, 405)

Because no similar formation is found on land, Darwin did not believe that atolls could have been raised from the seabed in a manner comparable to how mountains formed on land. Thus what is striking about Marsigli's and Darwin's approaches alike is how the comparative instinct imposed itself too fully on the emerging picture of the oceanic depths. As soundings were collated and pictures of the seabed constructed, these were forced within a known framework of overland geology. This exemplifies the period's difficulty in thinking about the ocean-space as an independent, geographical space, although new theories were emerging, breaking away from that approach.[5] Thus the geography of the ocean-space was still deeply rooted in a terrestrial approach that underpins geographical thought.

What I am emphasizing here is the notion that, until a bigger global picture was created and visualized sometime during the twentieth century, the vast enterprise of making sense of the ocean floor and its great depths was based on little more than land-centered guesswork. This is crucial with regard to this study, as it highlights a specific way of producing knowledge about the ocean-space. Certainly, through comparison with what was known and seen on land, conceptualizing the ocean depths' features had a starting point. However, structured interpretation of the sea's profile was to shift paradigms of global geology.

The key to comprehending the complex relation between knowing the sea floor and knowing the earth is in connecting the shape of the ocean to the shape of continents, rather than attempting to correlate specific features of either to the other. Certainly, the "shape of the oceans includes the shape of the edges of continents" (Laughton 2001, 97). Thus, while the similarities in shape between the coastlines of Africa and South America had been noticed, there was no framework available to understand the full scale of the observation. Furthermore, scientists were also collecting fossils and rock sediments that matched from either side of the ocean and overwhelmingly explained this by claiming that continents were once connected by land bridges (Wegener 1966, 1). This theory explained satisfactorily how fossils could, and indeed should, be similar on either side of the ocean. However, the land bridge theory acted as an obstacle to any scenario that would consider continental morphology. Thus making sense of the shape of continents required overturning conventional thought as well as asserting that continents had not always been in their present respective positions and having to provide an acceptable explanation for how that might have occurred.

In 1915, Arthur Wegener published *The Origin of Continents and Oceans*, a seminal book that outlined the theory of continental drift. In it, he "suggested that forces within the earth drove the continents as blocks drifting through the weaker oceanic crust" (Laughton 2001, 97). Wegener's theory was informed by "considering the map of the world, under the direct impression produced by the congruence of coastlines on either side of the Atlantic" (Wegener 1966, 1). In other words, it resulted from critical observation combined with examining available data. While its main tenet was not entirely new, Wegener's thesis is original in its thorough analysis of available data. Most significantly, however, this located his research firmly within a new framework of analysis. By looking at local phenomena in a global setting, Wegener sought to understand not only the earth's continental system but also how the continents themselves interacted among themselves and their subsequent effect on the ocean-space. This shift from one framework to another—that is, from a localized land-based explanation to a more complex, global one encompassing the oceans—provided the possibility for a vast theory of earth. Yet, this was not a straightforward process and indeed, Wegener himself was aware of the problems of working within a blinkered ideology, either a general scientific one or a more specifically disciplinary one. He writes:

> It is a strange fact, characteristic of the incomplete state of our present knowledge, that totally opposing conclusions are drawn about prehistoric conditions on our planet, depending on whether the problem is approached from the biological or the geophysical viewpoint.
>
> (Wegener 1966, 5)

One of the reasons cited for initial criticism of Wegener's theory was that he was a meteorologist and a geodesist, and thus was not seen as a credible scholar in matters of geology (King 1966, vii). Here the reticence toward Wegener's theory because he belonged to the wrong circles, albeit scientific circles, is redolent of these problems of trust in science and, following Mayhew (2005, 73), when considered against "the functioning of early modern European scientific communities and the ways in which they styled themselves," Wegener's exclusion from certain circles is unsurprising. Furthermore, Wegener and his theory were presumably also targeted by reactionary signs where "normal science (. . .) suppresses fundamental novelties because they are necessarily subversive of its basic commitments" (Kuhn 1962, 5). Certainly, Wegener's new theory of global geology constitutes a paradigm shift that necessitates a new framework of analysis.

The importance of the theory of continental drift, though misguided in its explanations, lies in the fact that it located the place of continents and the ocean within a global, interconnected discourse. King (1996, xxiv) states that:

> Wegener's pioneering work is not to be judged merely in terms of its correctness as a hypothesis of the earth's behaviour, although the greater part of his main contentions has surely been sustained by more recent findings; its greatest contribution has been a challenge which has stimulated enquiry and endeavour in so many fields of earth science.

In fact, as part of the discovery of the Atlantic Ocean, the Mid-Atlantic Ridge emerged as a feature that would challenge an immobile picture of the earth and instead posit that it might be mobile. By the time the link between the ridge and continental drift was discovered, it was the entire geography of the earth (oceanic and terrestrial) that had shifted frameworks. Henceforth, such physical features would not be considered in isolation but rather as part of a global network of phenomena. Following the arguments raised in Chapters 3 and 4 pertaining respectively to the cognitive process of discovery and the scientific measurement of the ocean-space, geographical knowledge production about the ocean-space is now located within a global framework. The story of *Bathybius haeckelii*, which will now be discussed, will develop this stance, bringing to the fore the manner in which knowing the ocean-space became a worldly enterprise.

5.3.2 *The case of* Bathybius haeckelii

The second example of placing the activity of sounding the ocean's depth within a larger epistemological context that will be discussed here considers the philosophical stance of the enterprise. As noted earlier, as well as sounding to ascertain the ocean's depth, the second principal aspect of fathoming was to obtain a sample of the seabed and determine its composition and what it might have to say about the earth and its history. Thus sounding was part of a larger desire to make sense of the earth and its origins, and also the origins of life itself. After Darwin's 1859 publication of *On the Origin of Species*, the scientific community was confronted with a plethora of questions that needed to be addressed before evolution provided a satisfactory philosophical backdrop. Indeed, while there was much excitement about the new theory of evolution, many questions were also brought to the fore, and nobody could yet offer convincing answers. In particular, now that there was no divine explanation to justify the origins of life, scientists were at pains to come up with their own scientific, provable explanation for the origins of life. Achieving this would not only provide Darwin's theory credibility but would also be a crowning achievement for science. Therefore, all eyes turned to the sea.

The idea that life had originated from the ocean was not a new one. Rehbock (1975, 505) writes that, in fact, it

> is one of the oldest speculations of natural history. Variations of this theme appear in the philosophy of the Milesian Anaximander and more recently in the speculative biology of the *Naturphilosophie* Lorenz Orken. Orken believed that life began as a primitive mucous substance which evolved from inorganic constituents existing in shallow waters.

Also, the fact that Charles Darwin had developed his theory of evolution by natural selection while at sea and while observing sea and island creatures helped to support this. Thus the ocean presented itself as the obvious place to look for these answers, and the seabed, essentially unknown, emerged as the likely provider of the required answers. Thus when Thomas Henry Huxley, known as "Darwin's Bulldog" for his staunch support of Darwin's theory of evolution, announced the discovery of an organism that bridged the gap between inorganic and organic matter, the scientific community celebrated the discovery for filling both biological and epistemological gaps in the new scientific paradigm (Kunzig 2000, 93).

Without delving into the history of cytology, the importance of Orken's theory is that it gave way to another concerning the contents of the cell and the nature of photoplasm, or animal cell substance. This was a popular subject of the time, and it was what Huxley was examining when he gave his paper "On Some Organisms Living at Great Depths in the North Atlantic Ocean" in 1868 (Huxley 1869). It is in this landmark paper that Huxley introduced *Bathybius haeckelii*, the apparently semi-live creature that appeared to be the link between inorganic and live matter (Figure 5.3).[6] This discovery stirred phenomenal interest, and the scientific community celebrated the discovery for filling biological and epistemological gaps of the new scientific paradigm. As Rozwadowski (2005, 163–164) writes:

> *Bathybius*'s arrival, awaited by geologists, cytologists, protozoologists, and evolutionists, whose lines of research had converged in the 1860s, fulfilled many scientific expectations. It provided a subcellar, elemental unit of life that served as the bridge between life and inorganic matter and promised to explicate Darwinian theory.

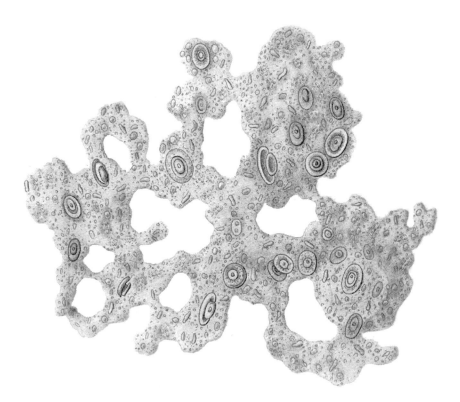

Figure 5.3 Drawing of *Bathybius haeckelii* by Haeckel, c. 1870.
Source: http://de.wikipedia.org/wiki/Bathybius

Bathybius thus gave Darwinism a piece that had, crucially, been missing and proved that science was able to give answers: it fitted within a model of monistic philosophy toward which "science had been striving in the nineteenth century" (Rehbock 1975, 522).

However, as it turned out, *Bathybius* did not exist. When HMS *Cyclops* sailed out in 1857, Huxley assigned the task of "obtaining, observing, and preserving whatever deposits might be brought up by the sounding device" (Rehbock 1975, 511). He specified for "the freshly brought up soundings [to be preserved] in a tolerably strong alcoholic mixture, so that the presence or absence of soft parts in them might be determined at any future time, under more convenient circumstances" (Huxley in Rehbock 1975, 511). When Huxley examined the ooze, he "saw it laced through with a network of slimy, gelatinous material that resembled egg-white, and that floated free, quivering, and shifting around in a lifelike way, when he shook the jar. Embedded in the slime were granules of unidentified material that the slime had apparently eaten" (Kunzig 2000, 94). Huxley (in Rehbock 1975, 516) thus wrote that he had discovered:

> deep-sea "Urschleim," which must, I think, be regarded as a new form of those simple animated beings which have recently been so well described by Haeckel in his "Monographie der Moneren."[7] I proposed to confer upon this new "Moner" the generic name of *Bathybius*, and to call it after the eminent Professor of Zoology in the University of Jena, *B. Haeckelii*.

However, in the end, *Bathybius* was "nothing more than the result of precipitation of calcium sulphate from the sea water in the deposit by alcohol in which it had been preserved" (Deacon 1997, 352). It was caused by the process that sought to examine it. It filled no ontological or epistemological gap. It was a precipitate that triggered too eager a response in an intellectual setting that, perhaps, too eagerly sought to make advances. As circumstances in various disciplines converged and made room for *Bathybius*, the scientific community saw what it wanted and needed to see to give credibility to the new Darwinian paradigm. Duly, *Bathybius* has been called "one of the most peculiar and fantastic errors ever committed in the name of science" and that it was "the product of an over-confident materialism, a vainglorious assumption that the secrets of life were about to be revealed" (Eiseley in Rehbock 1975, 505).

The story of *Bathybius* teases out the process of discovery and of knowledge production that, while in this particular instance brought on incorrect conclusions, nonetheless embodies the scientific method and spirit that characterized nineteenth-century science. Indeed, *Bathybius* was brought to life in an interdisciplinary scientific "intellectual environment congenial" to it and "however 'peculiar and fantastic' it might now appear, *Bathybius* was a rational construction, a snugly fitting piece in the intricate puzzle of Victorian science" (Rehbock 1975, 533). As a scientific story that was not only set at sea but which was also central to the understanding of the deep ocean-space from a cultural point of view, *Bathybius* is exemplary of a specific way of making knowledge about a geographical space. Certainly, through trying to connect a variety of disciplinary discourses, and solve philosophical and scientific problems while nonetheless remaining faithful to scientific methodology, the efforts put into understanding *Bathybius* and its global implications crucially highlight the problematics of the ocean-space and how it might be known scientifically. Considering both the shape of the oceans and the global presence of *Bathybius* within the perspective knowledge production, the significance of scales of space in which knowledge is made are brought to the fore.

5.3.3 *Scales of knowledge*

The questions of both plate tectonics and the *Bathybius* are deeply entrenched in issues of scale. In both instances, the globality of each phenomenon is essential to its nature: plate tectonics link every continent and ocean-space to each other through the earth, and *Bathybius*'s existence in all oceanic basins gave it credibility in debates about the origins of life. The practices and processes of making science on a ship are essential in relation to the stories discussed earlier. In the case of deep-sea soundings, local soundings became global through being inserted into a global frame of reference and considered in relation to other scientific fields. On the other hand, the initial ubiquity of the organism was an essential characteristic. However, this global scale was debunked within the setting of the laboratory. Thus, following Livingstone (2003, 12), "what is known, how knowledge is obtained, and the ways warrant is secured are all intimately bound up with the venues of science." Yet, ultimately, it was the locality of *Bathybius* that was its downfall: it was too local to the laboratory's test tube.

Furthermore, both global geology and *Bathybius* put emphasis on the process of the making of science and facts once the methodology has been ascertained. This particular facet of discovering facts about the ocean-space is heavily grounded within a wider, multi- and interdisciplinary approach to the earth as a whole. Through putting these oceanic discoveries against the background of cultural settings, the tight intertwinement between science and culture becomes apparent. Certainly, this all-encompassing way of looking at, studying, and seeking to understand the oceans exemplifies the nature of ocean science since the HMS *Challenger* expedition. What Wegener's theory and the short-lived *Bathybius* typify is knowledge that fits within wider, more complex pictures, moving beyond the intellectual ocean-space. In fact, the story of *Bathybius* further highlights the close intertwinement between cultural conditions and scientific settings, as scientists set out to find an organism that would fulfill a cultural as well as scientific role. This is central to the making of geographical knowledge. This is the cultural location of making knowledge, which, in the case of *Bathybius*, expressed itself through the mental space that created the right conditions for the invention of the creature. As such, the scientific study of the ocean-space emerges as a big science whose scales and practices typify a global project of constitutive knowledge. However, the making of geographical knowledge about the ocean-space was acted out on a variety of scales, including, uniquely, the floating laboratory of the ship: science and knowledge was also made locally.

5.4 Conclusion

Producing knowledge about the deep sea and the seabed was a complex, multilayered enterprise. In particular, technological, epistemological, cultural, and physical constraints converged and influenced the two examples discussed in this chapter. While the science of the Enlightenment contradicted gap-producing methods, the objective to know more about the ocean-space prevailed and drew maps that, though imprecise, led the way to more knowledge. Eventually, knowledge about the deep ocean-space was drawn on maps and the isometric lines of new bathymetric maps stood for what they represented, not a flawed methodology. Furthermore, the specific stories of plate tectonics and *Bathybius* are exemplary of the varied scales in which science took place. Through these two examples, the importance of considering the different locales of geography and the different scales upon which geographical knowledge is produced is put in evidence. As Kafka (in Carter 1999b, 295) put it, "the history of mankind is the instant between two strides taken by a traveller." In terms of knowing the deep ocean-space, understanding the gap between soundings and the true nature of deep-sea samples represent the history of knowing the ocean-space as a whole.

Notes

1 Using sound waves to measure the ocean's depth is the principle of echo sounding, which was invented in 1914 by Alexander Behm as a means to detect icebergs (Behm 1921).
2 The idea of the coast or the seabed as an intermediate space saying things about the phenomena found on either side of it is akin to Lévi-Strauss's explanation for the use of the coyote in Native American mythology. Lévi-Strauss writes: "The coyote (a carrion-eater) is an intermediary between herbivores and carnivores *like* mist between the sky and the earth; *like* the scalp between war and agriculture (the scalp is a war "crop"); like corn smut between wild and cultivated plants (it grows on the latter like the former); *like* clothing between "nature" and "culture"; *like* trash between the village and the wild; *like* ashes (and soot) between the hearth (on the ground) and ceiling (image of the celestial vault)" (Lévi-Strauss 1957, 259). I include this here to point out the significance of the concept of in-between. As such spaces, the coast and the seabed thus embody both their physical reality and the cultural no-man's land.
3 However, unlike the coast, the sea floor does not drastically change appearance with tides on a daily or twice daily basis. It does, though, exhibit significant change over geological time as seabed spreading changes the shape of oceans. I discuss this in Chapter 6.
4 Note: 1 fathom = 6 feet. The measure was originally defined as the distance of outstretched arms.
5 One such was George Darwin's (Charles's son) suggestion that oceanic basins "were the result of the gap left after the breakaway of the moon" (Laughton 2001, 95). There was no evidence to support this claim, and it was rapidly dismissed, though it is noteworthy in that it is not based on a terrestrial comparison.
6 Henceforth, I will refer to *Bathybius haeckelii* only as *Bathybius*.
7 Urschleim in "the most primitive possible form of life, pure protoplasm," which had been theorized by Ernst Haeckel (Kunzig 2000, 94).

6 Conclusion

In the previous chapters, this book has explored the production of knowledge about the ocean-space itself from the Age of Discovery to the late-nineteenth century, discussing events, ideas, and needs that shaped the discovery of the world's oceanic spaces. Gradually, the ocean-space was placed, delimited, and sounded, and its material and spatial properties began to be understood. At the same time, manners in which the ocean-space was utilized changed, be it through the adoption of new sea routes, the notion that seaside holidays had health benefits, or evolving fisheries techniques. These are evidence of a more utilitarian relationship with the ocean-space, which, while it still influences knowledge production, irrevocably shaped interactions with the ocean-space since the beginning of the twentieth century. Indeed, as technology began facilitating larger-scale retrieval of resources, such as fish and then hydrocarbons, the study of the ocean-space began to more specifically include that of its resources. This shift in interest is reflected in scholarly study through international conferences, such as one in 1883 in London and the founding of a number of research institutes, such as the Stazione Zoologica in Naples, Italy (1872); the Station Biologique in Roscoff, France (1872); the Marine Biological Association of the United Kingdom in Plymouth (1884); and the Marine Biological Laboratory in Woods Hole, Massachusetts, USA 1888. From then on, the study of the ocean-space would continue to move from a geographical to a biological and oceanographical subject area.

It therefore seems appropriate to bring this book to a conclusion by noting that, over the course of the last five years (during which time this book was written), through the effects of plate tectonics, the Atlantic Ocean has widened by approximately ten centimeters. At the current rate of sea floor spreading, by the end of the twenty-first century, Europe and Africa will be about 2.5 meters further away from the Americas than they are today. Thus in Iceland and other islands divided by the Mid-Atlantic Ridge, land that is presently over one meter inland will be eroding into the Atlantic Ocean within one hundred years. The physical geography of the Atlantic Ocean is changing as its absolute ocean-space is increasing. But, returning to the book's early dichotomy, over the same period, my own Atlantic Ocean has also changed, shaped by experiences on it and near it. Sailing past Start Point, where the channel becomes the Atlantic Ocean, and Lizard Point, Britain's most southerly promontory, or sounding creeks to find out whether they are safe overnight harbors are events that are part of *my* Atlantic ocean-space. These two ways of experiencing the Atlantic ocean-space are representative of different ways of understanding the geographical ocean-space. This chasm that has underpinned this thesis is described by Bernard Moitessier (2006, 268) in these words:

> [The] geography of the sailor is not always the one of the cartographer, for whom a cape is a cape with its longitude and latitude. For the sailor, a great cape is both very simple and extremely complex, with rocks, currents, furling seas, beautiful oceans, good winds and gusts, moments of happiness and of fright, fatigue, dreams, aching hands, an empty stomach, marvelous minutes and sometimes suffering. A great cape, for us, cannot be translated only into a latitude and a longitude. A great cape has a soul, with shadows and colors, very soft, very violent. A soul as smooth as that of a child, as hard as that of a criminal.

In this quote, numbers and experiences are superimposed to provide a fuller understanding of the Atlantic Ocean.

Whereas there will always be a discrepancy between various approaches to making sense of space, this book has demonstrated that a diachronic perspective that seeks to encompass a variety of elements and points of view is both possible and helpful. The broad, historical approach adopted was underpinned by a grand narrative of knowledge production. Over the period of this study, this methodological framework helped bring to the fore the multiplicity of ways in which scientific knowledge about the ocean-space was made across spaces, periods, and social strata, and highlight specific mechanisms at play in making knowledge about space. The interpretation and contextualization of each small story discussed in this book, and indeed this book's contribution to scholarship, is contingent upon locating them within a broad reconceptualization of the history of geographical knowledge as well as within the broad narrative of scientific knowledge production about the ocean-space.

In Chapter 2, a broad focus on the historical geographies helped us understand geographical knowledge production about faraway spaces built on the concept of tropicality as a lens. Specifically, examining the role of science in making knowledge travel called attention to the mechanisms of knowledge production through standardization as well as by the use of networks of trust. By then grounding the discussion within the locale of the ship as a scientific instrument and a space of science, different facets of science were brought together within a particular space in which science can occur. In the latter half of this chapter, the issues raised previously were considered in relation to the ocean-space, reflecting on how, to a certain extent, the ocean-space traversed consciousness as a historical and geographical void. Yet the usefulness of "historicizing" and "geographicalizing" the ocean-space was put in evidence as a manner of offering a new, varied perspective. Certainly, through examining the ocean-space on these terms, the ocean-space emerged as one that is in fact constructed scientifically as well as culturally, and whose place in geographical thought is diverse and shaped by a multiplicity of factors. Thus Chapter 2 established that considering the ocean-space in absolute, relative, or other single-minded terms fails to make sense of it as a geographical whole, both physically and perceptually, but also geographically, historically, and culturally.

Chapters 3, 4, and 5 built on this stance and, through the analysis of a range of archival materials, demonstrated specific ways in which knowing the Atlantic ocean-space is in fact both a scientifically complex and culturally intertwined enterprise. Chapter 3 discussed the discovery of the limits of the Atlantic Ocean and considered it from a cultural standpoint as well as within a cognitive framework. The question of the mental discovery of the Atlantic ocean-space brought to the fore the processes of knowledge production. Furthermore, this positioned the mental discovery of the Atlantic Ocean against the European paradigm shift that saw the basis of knowledge become science rather than biblically inspired. Chapter 4 highlighted the intertwined nature of scientific knowledge through the cultural and social settings of the period by examining the Enlightenment enterprise of mapping the World Ocean and focusing on nonexistent islands, the invention of the chronometer, and the measuring of magnetic fields at sea. Science, while preferred as a method to make knowledge, certainly emerged as indubitably affected by cultural factors. In particular, questions of how knowledge travels and how facts about the ocean-space are produced were exemplified with regard to nonexistent islands and the European discovery of the Gulf Stream. This showed how the strained relationship between different social circles and even within social networks (such as the Royal Navy or groups of fishermen) were influential in shaping knowledge about the ocean-space. The work of Maury on large-scale water movements provided a useful case study to explore various communication networks upon which he was able to draw and that gave legitimacy to his work in the eyes of sailors and merchants. Conversely, similar issues were seen with regard to Franklin's discussion of the Gulf Stream, though in this case to highlight a communication failure, since the Royal Navy did not believe the knowledge from the network of American fishermen. Halley's expedition aboard HMS *Paramore*, highlighted problematics entailed by scientists aboard ships not belonging to the Royal Navy network, something that was discussed again with regard to the HMS *Challenger* expedition.

Chapter 5 considered the globalization and scientific contextualization of ocean science. The discussion was located around the discovery of the Mid-Atlantic Ridge and the story of *Bathybius haeckelii* and demonstrated how knowledge of the Atlantic Ocean was, again, highlighted as both scientifically grounded and culturally located. In particular, these two stories were considered with regard to global scientific discoveries, putting emphasis on the importance of examining knowledge production on a large scale. By considering locales of science and discussing geographical discoveries simultaneously against the settings of local and global scales, I highlighted the importance of both perspectives when studying knowledge production about the Atlantic ocean-space. Throughout this final chapter, the challenges of negotiating both small-scale stories and grand narratives are particularly visible, in part because, by the end of the nineteenth century, scientific knowledge of the ocean-space became increasingly enmeshed with global geology, thus moving to an even broader outlook. Whereas the small stories of both *Bathybius* and Wegener's continental drift theory had previously been told within the history of science exemplifying that "the standard model for historicizing science is to locate specific pieces of work on as tight a context as possible, binding them ineluctably to the conditions of their production" (Secord 2004, 657), in the present context, they emerged as significant beyond the local conditions of their production and are important in relation to all the other stories discussed elsewhere in this book.

As a whole, this book was driven by an unconventionally broad methodological approach, which connects a number of small stories around a general, epistemological theme. By examining how the ocean-space has been negotiated as a geographical space through a sequence of scientifically driven enterprises, this historical perspective highlights the complexities of making sense of knowledge production diachronically. Archival material was interpreted, not each and individually in-depth, but rather against a conceptual "screen," to return to Driver and Martins' term for the conceptualization of the tropics (Driver and Martins 2005b, 5). Indeed, like amalgamated geographical imagination of the tropics, the geographically and scientifically known ocean-space hinges upon negotiating stories at a variety of scales, both geographical and historical, and interpreting them in light of an understanding of scientific knowledge.

However, the choice to focus on large-scale, scientific perspectives means that certain aspects of knowledge production about the ocean-space specifically and space in general are not included here, considering only specific economic or cultural trends in relation to scientific knowledge production. Thus Maury's economic interpretations were relevant, but this book did not develop the heavy economic influences, specifically those linked to the slave trade, on mapping the Atlantic Ocean and the Triangular Trade routes (Fanon 1952 and 1961; Gilroy 1993; Bhabha 2004). Certainly, the economic aspect of Atlantic Ocean historiography and its human implications underpin the human component of knowledge of the ocean-space. Similarly, the idea of the ocean-space as crucial within discourses of empire as a link between the core and the periphery was, some have argued, essential to the material workings of European empires (Steinberg 2009). Analyzing the entwinement between the ocean-space, empire, power, and identity would assuredly foreground an interesting set of transoceanic interactions, though a specifically imperial interpretation of the ocean-space would not add to the scientific knowledge of the geographical ocean-space.

Similarly, the cultural expressions of the ocean-space in literature and art were only reviewed briefly, primarily to demonstrate that the ocean's place in literature is more often as a backdrop than a protagonist or an active setting. However, a more detailed study of the ocean-space within the setting of literature and art would indicate how the cultural sphere produced knowledge about the ocean-space at different times. Certainly, grasping the broad dichotomy between science and culture has been crucial throughout this thesis, and by examining a selection of literary and artistic sources, the ocean-space could be located within a cultural narrative. This is the opinion of Klein (2002b, 3), who argues that the sea in literature "offers perhaps the best yardstick to assess the changing cultural conceptions of the sea across the centuries." Yet, again, while exploring this particular trope more in detail would have highlighted certain interesting links between various cultural spheres, this thesis was driven by scientific enterprises and thus considered cultural outputs in light of scientific discourse. However, a future project might build on this and Klein's work to locate oceanic cultural discourses and make sense of the ocean as a cultural and culturally rich space.

Certainly, focusing on different tropes to analyze the ocean-space would contribute to scholarship about the ocean-space in interesting ways. However, the focus here on science allowed consideration of the ocean-space first and foremost as a geographical space rather than a historical site. Further, over the extended period examined here, science, despite all the forms that it took (as discussed throughout this work), nonetheless remained closely tied to knowledge production. Thus as a constant against which to diachronically study knowledge production about the Atlantic ocean-space, science was more helpful than any other trope. Yet an attempt at being more inclusive of tangential sociocultural tropes might be a starting point for future, more all-encompassing studies that can explore the connections between various strands against a grand narrative.

Furthermore, this book is concerned with the ocean-space, considering specific global issues, such as large-scale water movements and the development of a global model for geology. These particular instances hint toward the possibility of a more global way of thinking about the ocean-space and the earth. Specifically, this book seeks to consolidate the idea that the World Ocean should be considered more as a unit than as composed of individual oceans. The establishment of the "continuous sea passages from ocean to ocean round the world" dates to the Age of Exploration, so it is surely possible to absorb this notion in future studies. And indeed, this book moves from considered ideas in development, placing borders on a worldly grid and mapping movements within and beyond it, toward finally relocating the ocean-space within a system of global geology and biology, thus working toward understanding how the ocean-space can be known as a whole and in relation to the World Ocean and the earth itself.

In effect, the work achieved in this book is principally twofold. First, it's historically long and geographically broad perspective of scientific knowledge production offered an alternative to those approaches that currently dominate historical and historical-geographical work. Yet this broad approach was reliant on much previous work and was never entirely divorced from the local stories from which it sought to distance itself. Indeed, these were crucial to the telling of this historical geography of the Atlantic Ocean. Second, by focusing on the ocean as a space worthy of the attention of human geographers, rather than physical geographers or maritime historians, it demonstrated that a truly oceanic perspective, which considers the ocean-space unopposed to land-based perspectives or processes, is both possible and useful. Yet examining the ocean-space from a primarily European perspective over a period of four centuries focusing on scientific knowledge production emerges as, rather than an end in itself, a way of segueing into a more comprehensive outlook on the earth and its geography. From this point of view, this book is but a stepping-stone toward further work, which would consider the place of the ocean-space in light of multiple physical and human interactions with it, on it, and inside it. Certainly, the ocean-space emerges as not only a location but also as a site of complex physical and human geographies, which, within a wide framework, is reinvented as a vibrant and exciting space.

Bibliography

Adorno, T. and Horkheimer, M. (2002) *Dialectic of Enlightenment: Philosophical Fragments*. Schmid Noerr, G. (Ed.) Jephcott, E. (Trans.). Stanford, CA: Stanford University Press.

Akerman, J. (Ed.) (2006a) *Cartographies of Travel and Navigation*. Chicago, IL: University of Chicago Press.

Akerman, J. (2006b) "Introduction" in Akerman (Ed.) (2006a): 1–15.

Alexander VI (1493) "Inter Caetera" Web document: http://www.catholic-forum.com. Last accessed 14/10/2008.

Allen, D. (2002) "The British Navy rules: Monitoring and incompatible incentives in the age of fighting sail" *Explorations in Economic History*, 39(2): 204–231.

Althusser, L. (1968) "Ideology and Ideological State Apparatuses" in Rivkin and Ryan (Eds.) (1998): 294–304.

Anderson, B. (2006) *Imagined Communities: Reflections on the Origin and Spread of Nationalism*. Revised edition. London: Verso.

Andrews, J. H. (1996) "Definitions of the Word 'Map', 1649–1996" Web document: http://www.usm.maine.edu/~maps/essays/andrews.htm. Last accessed 03/06/2007.

Armitage, D. (2002) "Three Concepts of Atlantic History" in Armitage and Braddick (Eds.) (2002): 11–27.

Armitage, D. and Braddick, M. J. (Eds.) (2002a) *The British Atlantic World, 1500–1800*. Basingstoke: Palgrave.

Armitage, D. and Braddick, M. J. (2002b) "Introduction" in Armitage and Braddick (Eds.) (2002a): 1–7.

Arnold, D. (1998) "India's place in the Tropical World, 1770–1930" *The Journal of Imperial and Commonwealth History*, 26(1): 1–21.

Arnold, D. (2000) "'Illusory Riches': Representations of the Tropical World, 1840–1950" *Singapore Journal of Tropical Geography*, 21(1): 6–18.

Arnold, D. (2005) "Envisioning the Tropics: Joseph Hooker in India and the Himalayas, 1848–1850" in Driver and Martins (Eds.) (2005a): 137–155.

Augé, M. (1995) *Non-Places: Introduction to an Anthropology of Supermodernity*. Howe, J. (Trans.). London: Verso.

Bachelard, G. (2005) *La philosophie du non*. Paris: press Universitaire de France.

Bailyn, B. (2002) "Preface" in Armitage and Braddick (Eds.) (2002a): xiv–xx.

Bailyn, B. (2005) *Atlantic History: Concept and Contours*. London: Harvard University Press.

Barber, P. (Ed.) (2005) *The Map Book*. London: Weidenfeld and Nicholson.

Barney, S., Beach, J. A., Berghof, O. and Lewis, W. J. (Eds.) (2006) *The Etymologies of Isidore of Seville*. Cambridge: Cambridge University Press.

Barry, A. (1993) "A history of measurement and the engineers of space" *The British Journal for the History of Science*, 26(4): 459–468.

Bartky, I. (1989) "The adoption of Standard Time" *Technology and Culture*, 30(1): 25–56.

Bassin, M. (1999) "History and philosophy of geography" *Progress in Human Geography*, 23(1): 109–117.

BBC News (2007) "Russia Plants Flag under N Pole—02/08/2007" Web document: http://news.bbc.co.uk/1/hi/world/europe/6927395.stm. Last accessed 05/06/2009.

BBC News (2008a) "Australia Extends Rights Over Sea—21/04/2008" Web document: http://news. bbc.co.uk/1/hi/world/Asia-pacific/7358432.stm. Last accessed 05/06/2009.

BBC News (2008b) "UK Makes Atlantic Sea Bed Claim—27/08/2008" Web document: http://news. bbc.co.uk/1/hi/uk/7583353.stm. Last accessed 05/06/2009.

BBC History (2009) "Captain Cook: Explorer, Navigator and Pioneer" Web document: http://www.bbc.co.uk/history/british/empire_seapower/captaincook_01.shtml. Last accessed 07/05/2009.

Behm, A. (1921) "Means for sounding or measuring distance in water" Patent application 1,649,378 submitted to the United States Patent Office 7 July, 1921.

Bentley, J. (1996) "Cross-cultural interaction and periodization in world history" *The America historical Review*, 101(3): 749–770.

Bentley, J. (1999) "Sea and ocean basins as frameworks of historical analysis" *Geographical Review*, 89(2): 215–224.

Bergson, H. (2007) *Essai sur les données immédiates de la conscience*. Paris: presses Universitaires de France.

Berthon, S. and Robinson, A. (1991) *The Shape of the World: The Mapping and Discovery of the Earth*. London: George Philip.

Bhabha, H. K. (2004) *The Location of Culture*. London: Routledge.

Bird, A. (2007) "What is scientific progress?" *Noûs*, 41(1): 64–89.

Black, J. (2003) *Visions of the World: A History of Maps*. London: Mitchell Beazley.

Blank, P. W. (1999) "The Pacific: A Mediterranean in the making?" *Geographical Review*, 89(2): 265–277.

Bose, S. (2006) *A Hundred Horizons: The Indian Ocean in the Age of Global Empire*. Cambridge, MA: Harvard University Press.

Braudel, F. (2001) *The Mediterranean in the Ancient World*. London: Penguin.

Bravo, M. (2006) "Geographies of exploration and improvement: William Scoresby and Arctic Whaling, 1782–1822" *Journal of Historical Geography*, 32: 512–538.

Broome, R. (2002) *Amerike: The Briton Who Gave America its Name*. Stroud: Sutton Publishing Ltd.

Burney, J. (1803–1817) *Chronological History of the Discoveries in the South Sea or Pacific Ocean*. London: G. & W. Nicols.

Burstyn, H. (1975) "Science pays off: Sir John Murray and the Christmas Island Phosphate Industry, 1886–1914" *Social Studies of Science*, 5(1): 5–34.

Burstyn, H. (2001) "'Big science' in Victorian Britain: The Challenger Expedition (1872–6) and Its Report (1881–95)" in Deacon, Rice and Summerhayes (Eds.) (2001): 49–55.

Canguilhem, G. (1980) *La connaissance de la vie*. 2nd edition. Paris: Librairie Philosophique J. Vrin.

Canguilhem, G. (1981) *Idéologie et rationalité dans l'histoire des sciences de la vie: Nouvelles études d'histoire et le philosophie des sciences*. 2nd edition. Paris: Librairie Philosophique J. Vrin.

Canny, N. (2001) "Atlantic history: What and why?" *European Review*, 9(4): 399–411.

Carter, P. (1999a) "Dark with Excess of Bright: Mapping the Coastlines of Knowledge" in Cosgrove (Ed.) (1999): 125–147.

Carter, P. (1999b) "Gaps in Knowledge: The Geography of Human Reason" in Livingstone and Withers (Eds.) (1999a): 295–318.

Cavell, R. (2005) "Geographical immediations: Locating 'The English Patient'" *New Formations*, 57: 95–105.

Central Intelligence Agency (CIA) (2008) "The 2008 World Factbook" Web document: http://www.cia.gov/library/publications/the-world-factbook/index. html. Last accessed 26/05/2009.

Césaire, A. (1969) *Une tempête: d'après "La Tempête" de Shakespeare*. Paris: Seuil.

Chang, H. (2004) *Inventing Temperature: Measurement and Scientific Progress*. Oxford: Oxford University Press.

Charette, M. and Smith, W. H. F. (2010) "The volume of the Earth's Ocean" *Oceanography*, 23(2): 112–114.

Chaudhuri, K. N. (1985) *Trade and Civilisation in the Indian Ocean: An Economic History from the Rise of Islam to 1750*. Cambridge: Cambridge University Press.

Chaunu, H. and Chaunu, P. (1955–59) *Séville et l'Atlantique*. 8 volumes. Paris: Armand Colin and SeVpeM.

Colapinto, J. (2008) "Secrets of the Deep: The dispute over sunken treasure" *The New Yorker*, 84(8): 44–55.

Collingwood, R. G. (1993) *The Idea of History with Lectures 1926–1928*. Dussen, J. (Ed.). Oxford: Oxford University Press.

Collingwood, V. (2003) *Captain Cook: The Life, Death and Legacy of History's Greatest Explorer*. London: Ebury Press.

Conan Doyle, A. (2004) *The Sign of Four*. Fairfield, IA: 1st World Publishing.

Connery, C. (2006) "*There Was No More Sea*: The suppression of the ocean, from the bible to cyberspace" *Journal of Historical Geography*, 32: 494–511.

Cook, A. (2006) "Surveying the Seas: Establishing the Sea Route to the East Indies" in Akerman (Ed.) (2006): 69–96.

Cook, H. and Lux, D. (1998) "Closed circles or open networks?: Communicating at a distance during the scientific revolution" *History of Science*, 36: 179–211.

Cooke, M. (1999) "Mediterranean thinking: From Netizen to Medizen" *Geographical Review*, 89(2): 290–300.

Corbin, A. (1994) *The Lure of the Sea: The Discovery of the Seaside in the Western World 1750–1840*. Phelps, J. (Trans.). Cambridge: Polity Press.

Corfield, R. (2003) *The Silent Landscape: The Scientific Voyage of HMS Challenger*. Washington, DC: Joseph Henry Press.

Cosgrove, D. (1994) "Contested global visions: One-world, whole earth, and the Apollo Space photographs" *Annals of the Association of American Geographers*, 84(2): 270–294.

Cosgrove, D. (Ed.) (1999). *Mappings*. London: Reaktion Books.

Cosgrove, D. (2005) "Apollo's Eye: A Cultural Geography of the Globe" Hettner lecture at Universität heidelberg, Germany.

Cosgrove, D. (2008) *Geography and Vision: Seeing, Imagining and Representing the World*. London: I. B. Tauris.

Culler, J. (1975) "The Linguistic Foundation" in Rivkin and Ryan (Eds.) (1998): 73–75.

Cunningham, G. W. (1914) "Bergson's concept of duration" *The Philosophical Review*, 23(5): 525–539.

Darwin, C. (1958) *The Voyage of the Beagle*. New York, NY: Bantam.

Davidson, P. (2005) *The Idea of North*. London: Reaktion.

Davos, C. (1999) "On determining the social relevance of oceanography" *Progress in Oceanography*, 44: 457–468.

Dawber, R. (1969) "Fingernail growth in normal and Psoriatic subjects" *British Journal of Dermatology*, 82(5): 454–457.

Dawson, N.-M. (2000) *L'atelier Delisle: l'Amérique du Nord sur la table à dessin*. Sillery, QC: edition du Septentrion.

Deacon, M. (1997) *Scientists and the Sea, 1650–1900: A Study of Marine Science*. 2nd edition. Aldershot: Ashgate.

Deacon, M., Rice, T. and Summerhayes, C. (Eds.) (2001a) *Understanding the Oceans: A Century of Ocean Exploration*. London: UCL Press.

Deacon, M. and Summerhayes, C. (2001b) "Introduction" in Deacon, Rice and Summerhayes (2001a): 1–23.

Deniau, J.-F. (2002) *Dictionnaire amoureux de la mer et de l'aventure*. Paris, France: Plon.

Dening, G. (2004) "Deep Times, Deep Spaces: Civilizing the Sea" in Klein and Mackenthun (Eds.) (2004): 13–35.

Denis, D. and Friendly, M. (2001) "Milestones in the History of Thematic Cartography, Statistical Graphics, and Data Visualization" Web document: http:// www.math.yorku. ca/SCS/Gallery/Milestone. Last accessed 02/10/2013.

Descartes, R. (1991) *"Discours de la méthode" suivi de "La Dioptrique"* Paris: Gallimard.

Driver, F. (1988) "The historicity of human geography" *Progress in Human Geography*, 12(4): 497–506.

Driver, F. (2004a) "Imagining the Tropics: Views and visions of the tropical world" *Singapore Journal of Tropical Geography*, 25(1): 1–17.

Driver, F. (2004b) "Distance and disturbance: Travel, exploration and knowledge in the nineteenth century" *Transactions of the Royal Historical Society*, 14: 73–92.

Driver, F. and Martins, L. (2002) "John Septimus Roe and the art of navigation; c. 1815–1830" *History Workshop Journal*, 54: 144–157.

Driver, F. and Martins, L. (Eds.) (2005a) *Tropical Visions in and Age of Empire*. London: Chicago University Press.

Driver, F. and Martins, L. (2005b) "Views and Visions of the Tropical World" in Driver and Martins (Eds.) (2005a): 3–20.

Driver, F. and Martins, L. (2006) "Shipwreck and salvage in the tropics: the case of HMS *Thetis*, 1830–1854" *Journal of Historical Geography*, 32: 539–562.

Driver, F., Livingstone, D. and Thrift, N. (1995) "The geography of truth" *Environment and Planning D: Society and Space*, 13(1): 1–3.

Duncan, J. (2000) "The struggle to be temperate: Climate and "moral masculinity" in Mid-nineteenth century Ceylon" *Singapore Journal of Tropical Geography*, 21(1): 34–47.

Earle, S. (2005) "Foreword" in Rozwadowski (Ed.) (2005): ix–xii.

Edney, M. (1999) "Reconsidering Enlightenment Geography and Map Making: Reconnaissance, Mapping, Archive" in Livingstone and Withers (Eds.) (1999a): 165–198.

Ehrenberg, R. (2006) *Mapping the World: An Illustrated History of Cartography*. Washington, DC: National Geographic Society.

Elliot, J. (1990) "The Seizure of Overseas Territories by the European Powers" in Pohl, H. (Ed.) (1990): 43–61.

Evans, E. (1996) *Ireland and the Atlantic Heritage: Selected Writings*. Dublin (Ireland): Lilliput Press.

Fanon, F. (1952) *Peau noire, masques blancs*. Paris: Seuil.

Fanon, F. (1961) *Les damnés de la terre*. Paris, France: Découverte.

Field, K. (2005) "Maps still matter—don't they?" *The Cartographic Journal*, 42(2): 81–82.

Food and Agriculture Organization (2009) "Fisheries and Aquaculture in Our Changing Climate" Policy Brief of the FAO for the UNFCCC COP-15.

Foucault, M. (2001) *Histoire de la folie à l'âge classique*. Paris, France: Gallimard.

Foucault, M. (2004) *L'archéologie du savoir*. Paris, France: Gallimard.

Francis, J. M. (Ed.) (2006) *Iberia and the Americas: Culture, Politics and History*. Oxford: Abc-Clio.

Franklin, B. (1806) *The Complete Works, in Philosophy, Politics and Morals of the Late Dr. Benjamin Franklin, Now First Collected and Arranges with Memoirs of His Early Life* Vol. 2. London: Longman, Hurst, Rees and Orme.

Friendly, M. (2009) "Milestones in the History of Thematic Cartography, Statistical Graphics, and Data Visualization" Web document: http://www.math.yorku.ca/ SCS/Gallery/milestone/milestone.pdf. Last accessed 10/01/2014.

Gieryn, T. (2002) "Three truth spots" *Journal of History of the Behavioral Sciences*, 38(2): 113–132.

Gigerenzer, G. (1991) "From tools to theories: A heuristic of discovery in cognitive psychology" *Psychological Review*, 98(2): 254–267.

Gilroy, P. (1993) *The Black Atlantic: Modernity and Double Consciousness*. London: Verso.

Godlewska, A. M. C. (1999) "From Enlightenment Vision to Modern Science? Humboldt's Visual Thinking" in Livingstone and Withers (Eds.) (1999a): 236–279.

Graham, B. and Nash, C. (Eds.) (2000) *Modern Historical Geographies*. Harlow: Prentice Hall.

Green, W. (1995) "Periodizing world history" *History and Theory*, 34(2): 99–111.

Greppi, C. (2005) " 'On the Spot': Traveling Artists and the Iconographic Inventory of the World, 1769–1859" in Driver and Martins (Eds.) (2005a): 23–42.

Guha, R. and Spivak, G. (Eds.) (1988) *Selected Subaltern Studies*. Oxford: Oxford University Press.

Hale, Jr. E. (1854) "Maury's sailing directions: Commerce of the United States" *Hunt's Merchants' Magazine*, 30: 531–547.

Harland-Jacobs, J. (1999) " 'Hands across the sea': The Masonic network, British imperialism, and the North Atlantic world" *Geographical Review*, 89(2): 237–253.

Harris, S. (1998) "Long-distance corporations, big sciences, and the geography of knowledge" *Configurations*, 6(2): 269–304.

Harvey, D. C. and Mark Riley (2005) "Landscape archaeology, heritage and the community in Devon: An oral history approach" *International Journal of Heritage Studies*, 11(4): 265–288.

Heffernan, M. (1999) "Historical Geographies of the Future: Three Perspectives from France, 1750–1825" in Livingstone and Withers (Eds.) (1999a): 125–164.

Herschel, J. F. W. (Ed.) (1849) *A Manual of Scientific Enquiry, Prepared for the Use of Her Majesty's Navy and Adapted for Travellers in General*. London: John Murray.

Heuninck, T. (Ed.) (2005) *Le Cours des Glénans*. 6th edition. Paris, France: Seuil.

History of Marine Animal Populations (HMAP) (2011) "HMAP Data Pages" Web document: http:// www.hull.ac.uk/hmap/. Last accessed 13/10/2011.

Hoagland, P., Jin, D. and Kite-Powell, H. (2003) "The optimal allocation of ocean space: Aquaculture and wild-harvest fisheries" *Marine Resource Economics*, 18(2): 129–147.

Holdore, L. and May, W. E. (1973) *A History of Marine Navigation*. Henley-on-Thames: Foulis.

Holmes, D. (Ed.) (2005) *Britain and the Sea*. London: Royal Press.

Houghton, P. (1996) *People of the Great Ocean: Aspects of Human Biology of the Early Pacific*. Cambridge: Cambridge University Press.

Hourani, G. F. (1995) *Arab Seafaring in the Indian Ocean in Ancient and Early Medieval Times*. Revised and Expanded by Carswell, J. Princeton, NJ: University of Princeton Press.

Hunt, F. (1854) "Maury's sailing directions" *Hunt's Merchants' Magazine*, 30(5): 531–547.

Hurlbut, G. (1888) "The origin of the name 'America'" *Journal of the American Geographical Society of New York*, 20: 183–196.

Huxley, T. H. (1869) "On some organisms living at great depths in the North Atlantic Ocean" *Quarterly Journal of Microscopical Science*, 8: 203–212.

Imray, J. F. and Jenkins, H. D. (1884) "Atlantic Ocean Pilot: The Seaman's Guide to the Navigation of the Atlantic Ocean" followed by W. H. Rosser's "Notes on the Physical Geography of the Atlantic Ocean". London: Imray.

Ingold, T. (2000) *The Perception of the Environment: Essays in Livelihood, Dwelling and Skill*. London: Routledge.

International Court of Justice—Cour International de justice (ICJ-CIJ) (1953a) "Minquiers and Ecrehow Case: Summary of the Judgment of 17 November 1953" Web document: http://www.icj-cij.org/docket/files/17/2025.pdf. Last accessed 10/01/2014.

International Court of Justice—Cour Internationale de Justice (ICJ-CIJ) (1953b) "Affaire des Minquiers et des Ecréhous: Judgment of 17 November 1953" Web document: http://www.icj-cij.org/docket/files/17/2022.pdf. Last accesses 10/01/2014.

International Hydrographic Organization (IHO) (1953) "Limits of Oceans and Seas—Special Publication 23" Monte Carlo, Monaco: IHO.

Jacob, C. (1996) "Toward a cultural history of Cartography" *Imago Mundi*, 48: 191–198.

James Ford Bell Library (2009) "Welcome to the James Ford Bell Library" Web document: http://bell.lib.umn.edu/. Last accessed 27/04/2009.

Johns, W. E., Bane, J. M., Shay, T. J. and Watts, D. R. (1995) "Gulf Stream structure, transport, and recirculation near 68°W" *Journal of Geophysical Research*, 100(C1): 817–838.

Kant, I. (1995) "What Is Enlightenment?" in Kramnick (Ed.) (1995a): 1–7.

Kant, I. (2003) *Critique of Pure Reason*. Meiklejohn, J. M. D. (Trans.). Mineola, NY: Dover.

Kennedy, D. (1990) "The Perils of the Midday Sun: Climatic Anxieties in the Colonial Tropics" in MacKenzie (Ed.) (1990): 118–140.

Kern, S. (2001) *The Culture of Time and Space, 1880–1918*. Cambridge, MA: Harvard University Press.

King, B. C. (1966) "Introduction" in Wegener (1966): vii–xxiv.

Kirby, D. and Hinkkanen, M.-L. (2000) *The Baltic and the North Seas*. London: Routledge.

Klein, B. (Ed.) (2002a) *Fictions of the Sea: Critical Perspectives on the Ocean in British Literature and Culture*. Aldershot: Ashgate.

Klein, B. (2002b) "Introduction: Britain and the Sea" in Klein (Ed.) (2002a): 1–12.

Klein, B. and Mackenthun, G. (2000) *Call for Papers for the "Sea Changes; Historicizing the Ocean, c. 1500—c. 1900. Conference"* held at the University of Greifswald, Germany.

Klein, B. and Mackenthun, G. (Eds.) (2004) *Sea Changes: Historicizing the Ocean*. London: Routledge.

Koh, H. H. (1998) "Is International Law really State Law?" *Harvard Law Review*, 111(7): 1824–1861.

Kramnick, I. (Ed.) (1995a) *The Portable Enlightenment Reader*. London: Penguin.

Kramnick, I. (1995b) *"Introduction"* in Kramnick (Ed.) (1995a): ix–xxiii.

Kuhn, T. (1962) *The Structure of Scientific Revolutions*. Chicago, IL: University of Chicago Press.

Kuiper, K. (Ed.) (1995) *Merriam-Webster's Encyclopaedia of Literature*. Springfield, MA: Merriam-Webster.

Kunzig, R. (2000) *Mapping the Deep: The Extraordinary Story of Ocean Science*. London: Penguin.

Laloë, A.-F. (2009) *Knowing the Ocean-Space: An Atlantic Case Study*. School of Geography, University of Exeter. Unpublished doctoral thesis.

Lambert, D., Martins, L. and Ogborn, M. (2006) "Currents, visions and voyages: Historical geographies of the sea" *Journal of Historical Geography*, 32: 479–493.

Latour, B. (1986) "Visualisation and cognition: Thinking with hands and eyes" *Knowledge and Society*, 6: 1–10.

Latour, B. (1987) *Science in Action: How to Follow Scientists and Engineers through Society*. Cambridge, MA: Harvard University Press.

Latour, B. and Woolgar, S. (1986) *Laboratory Life: The Construction of Scientific Facts*. Princeton, NJ: Princeton University Press.

Laughton, A. (2001) "Shape as a key to understanding the geology of the oceans" in Deacon, Rice and Summerhayes (Eds.) (2001): 92–107.

Lavery, B. (1989) *Nelson's Navy: The Ships, Men, and Organization, 1793–1815*. London: Conway Maritime Press.

Lévi-Strauss, C. (1957) *Anthropologie Structurale*. Vol. 1. Paris, France: Pocket.

Lewis, M. W. and Wigen, K. E. (1997) *The Myth of Continents: A Critique of Metageography*. London: University of California Press.

Lewis, M. W. and Wigen, K. E. (1999) "A maritime response to the crisis in area studies" *The Geographical Review*, 89(2): 161–168.

Library of Congress (LoC) "Collection Highlights" Web document: http://www.loc.gov. Last accessed 28/04/2008.

Ligi, M., Bonatti, E., Bortoluzzi, G., Carrara, G., Fabretti, P., Gilod, D., Peyve, A., Skilotnev, S. and Turko, N. (1999) "Bouvet triple junction in the South Atlantic: Geology and evolution" *Journal of Geophysical Research*, 104(B12): 29365–29385.

Linebaugh, P. and Rediker, M. (2000) *The Many-Headed Hydra: Sailors, Slaves, Commoners, and the Hidden History of the Revolutionary Atlantic*. London: Verso.

Linklater, E. (1974) *The Voyage of the Challenger*. London: Cardinal.

Livingstone, D. N. (1992) *The Geographical Tradition: Episodes in the History of a Contested Enterprise*. Oxford: Blackwell.

Livingstone, D. N. (1999) "Geographical Inquiry, Rational Religion, and Moral Philosophy: Enlightenment Discourses on the Human Condition" in Livingstone and Withers (Eds.) (1999): 94–199.

Livingstone, D. N. (2000) "Tropical Hermeneutics: Fragments for a historical narrative" *Singapore Journal of Tropical Geography*, 21(1): 92–98.

Livingstone, D. N. (2003) *Putting Science in Its Place: Geographies of Scientific Knowledge*. London: University of Chicago Press.

Livingstone, D. N. and Withers, C. (Eds.) (1999a) *Geography and Enlightenment*. London: University of Chicago Press.

Livingstone, D. N. and Withers, C. (1999b) "Introduction: On Geography and Enlightenment" in Livingstone and Withers (Eds.) (1999a): 1–31.

Livingstone, D. N. and Withers, C. (1999c) "Preface and Acknowledgements" in Livingstone and Withers (Eds.) (1999a): vii–viii.

Lorimer, H. (2003) "Telling small stories: Spaces of knowledge and the practice of geography" *Transactions of the Institute of British Geographers*, 28(2): 197–217.

MacDonald, F. (2006) "The last outpost of Empire: Rockall and the Cold War" *Journal of Historical Geography*, 2: 627–647.

MacKenzie, J. (Ed.) (1990) *Imperialism and the Natural World*. Manchester: Manchester University Press.

Mancke, E. (1999) "Early modern expansion and the politicization of the oceanic space" *Geographical Review* 89(2): 225–236.

Martins, L. (1998) "Navigating in Tropical Waters: British maritime views in Rio de Janeiro" *Imago Mundi*, 50: 141–155.

Martins, L. (2000) "A naturalist's vision of the Tropics: Charles Darwin and the Brazilian landscape" *Singapore Journal of Tropical Geography*, 21(1): 19–33.

Maury, M. F. (1851) *Explanations and Sailing Directions to Accompany the Wind and Current Charts*. Washington, DC: US Navy.

Maury, M. F. (1855) *The Physical Geography of the Sea and Its Meteorology*. New York, NY: Harper.

Maury, M. F. (2003) *The Physical Geography of the Sea and Its Meteorology*. Mineola, NY: Dover.

Mayhew, R. (2005) "Mapping science's imagined community: Geography as a Republic of Letters, 1600–1800" *British Journal for the History of Science*, 38(1): 73–92.

Mayhew, S. (1997) *The Dictionary of Geography*. 2nd edition. Oxford: Oxford University Press.

McClintock, A. (1995) *Imperial Leather: Race, Gender and Sexuality in the Colonial Context*. London: Routledge.

McCook, S. (1996) "'It May Be Truth, but It Is Not Evidence': Paul du Chaillu and the legitimation of evidence in the Field Sciences" *Osiris*, 11 (2nd Series): 177–197.

Melville, H. (1963) *Moby Dick*. London: Dent.

Miller, D. P. (2002) "The 'Sobel Effect': The amazing tale of how multitudes of popular writers pinched all the Best Stories in the history of Science and became rich and famous while historians languished in accustomed poverty and obscurity, and how this transformed the world. A reflection on a Publishing Phenomenon" *Metascience*, 11(2): 185–200.

Mitchell, D. (2000) *Cultural Geography: A Critical Introduction*. London: Blackwell.

Moitessier, B. (2006) *La longue route: seul entre mers et ciels*. Paris, France: J'ai lu.

Monmonier, M. (2006) *From Squaw Tit to Whorehouse Meadow: How to Name, Claim and Inflame*. Chicago, IL: University of Chicago Press.

Morphy, F. and Morphy, H. (2006) "Tasting the waters: Discriminating identities in the waters of Blue Mud Bay" *Journal of Material Culture*, 11(1–2): 67–85.

Nash, C. (2000) "Historical Geographies of Modernity" in Graham and Nash (Eds.) (2000): 13–40.

National Oceanic and Atmospheric Administration (NOAA) (2009) "1785: Benjamin Franklin's 'Sundry Maritime Observations'" Web document: http://oceanexplorer.noaa.gov/history/readings/gulf/gulf.html. Last accessed 21/05/2009.

Naylor, S. (2000a) "'That very garden of South America': European surveyors in Paraguay" *Singapore Journal of Tropical Geography*, 21(1): 48–62.

Naylor, S. (2000b) "Spacing the can: Empire, modernity, and the globalisation of food" *Environment and Planning A*, 32: 1625–1639.

Naylor, S. (2002) "The field, the museum and the lecture hall: The spaces of natural history in Victorian Cornwall" *Transactions of the Institute of British Geographers*, 27: 494–513.

Naylor, S. (2005a) "Historical geographies of science: Places, contexts, cartographies" *British Journal for the History of Science*, 38(1): 1–12.

Naylor, S. (2005b) "Writing the region: Jonathan Couch and the Cornish fauna" *Interdisciplinary Science Reviews*, 30(1): 33–45.

Naylor, S. (2006) "Nationalizing provincial weather: Meteorology in nineteenth-century Cornwall" *British Journal for the History of Science*, 39(3): 407–433.

Naylor, S. (2008) "Historical geographies: Geographies and historiographies" *Progress in Human Geography*, 32(2): 265–274.

O'Beirne, E. (2006) "Mapping the *non-lieu* in Marc Augé's writings" *Forum for Modern Language Studies*, 42(1): 38–50.

Ogborn, M. (1998) "The capacities of the state: Charles Davenant and the management of the Excise, 1683–1698", *Journal of Historical Geography*, 24(3): 289–312.

Ogborn, M. (2000) "Historical geographies of globalisation, c. 1500–1800" in Graham and Nash (Eds.) (2000): 43–69.

Ogborn, M. (2002) "Writing travels: Power, knowledge and ritual on the English East India Company's early voyages" *Transactions of the Institute of British Geographers*, 27(2): 155–171.

Ogborn, M. (2004) "*Geographia*'s pen: Writing, geography and the arts of commerce, 1660–1760" *Journal of Historical Geography*, 30: 294–315.

Ogborn, M. (2005) "Atlantic Geographies" *Social and Cultural Geography*, 6(3): 379–385.

Oliver, D. (1989) *The Pacific Islands*. Honolulu, HI: University of Hawai'i Press.

Ophir, A. and Shapin, S. (1991) "The place of knowledge: A methodological survey" *Science in Context*, 4(1): 3–21.

Ordnance Survey (OS) (2008) *A Guide to Coordinate Systems in Great Britain*. Southampton: Ordnance Survey.

Organisation for Economic Co-operation and Development (OECD) (2008) *Space Technologies and Climate Change: Implications for Water Management, Marine Resources and Maritime Transport*. Paris, France: OECD.

Pagh, N. (2001) *At Home Afloat: Women on the Waters of the Pacific Northwest*. Alberta, AB: University of Calgary Press.

Parry, J. H. (1981) *The Discovery of the Sea*. London: University of California Press.

Pearson, M. (2003) *The Indian Ocean*. London: Routledge.

Philo, C. (1995) "Journey to asylum: A medical-geographical idea in historical context" *Journal of Historical Geography*, 21(2): 148–168.

Pickering, A. (1995) *The Mangle of Practice: Time, Agency and Science*. London: University of Chicago Press.

Pickles, J. (2004) *A History of Spaces: Cartographic Reason, Mapping and the Geo-Coded World*. London: Routledge.

Pirtle, C. (2000) "Military uses of ocean space and the law of the sea in the New Millennium" *Ocean Development & International Law*, 31(1): 7–45.

Pohl, H. (Ed.) (1990) *The European Discovery of the World and Its Economic Effects on Pre-Industrial Society, 1500–1800*. Stuttgart, Germany: Steiner.

Poincaré, H. (1908) *La Science et l'Hypothèse*. Paris, France: Flammarion.

Popper, K. (1959) *The Logic of Scientific Discovery*. London: Hutchinson.

Popper, K. (1994) *The Myth of the Framework: In Defence of Science and Rationality*. Notturno, M. A. (Ed.). London: Routledge.

Powell, R. (2007) "Geographies of science: Histories, localities, practices, futures" *Progress in Human Geography*, 31(3): 309–329.

Powell, R. (2008) "Becoming a geographical scientist: Oral histories of Arctic fieldwork" *Transactions of the Institute of British Geographers*, 33(4): 548–565.

Rehbock, P. (1975) "Huxley, Haeckel, and the Oceanographers: The case of *Bathybius haeckelii*" *Isis*, 66(4): 504–533.

Reidy, M. (2008) *Tides of History: Ocean Science and Her Majesty's Navy*. London: University of Chicago Press.

Rice, T. (2001) "The Challenger Expedition: The End of an Era or a New Beginning?" in Deacon, Rice and Summerhayes (Eds.) (2001): 27–48.

Richards, M. and Trinkausc, E. (2009) "Isotopic evidence for the diets of European Neanderthals and early modern humans" *Proceedings of the National Academy of Science*, 106(38): 16034–16039.

Rivkin, J. and Ryan, M. (Eds.) (1998) *Literary Theory: An Anthology*. Revised Edition. London: Blackwell.

Roach, J. (1996) *Cities of the Dead: Circum-Atlantic Performance*. New York, NY: Columbia University Press.

Rose, G. A. (2007) *Cod: The Ecological History of the North Atlantic Fisheries*. St John's, NL: Breakwater Books Ltd.

Rose, M. (2006) "Gathering 'dreams of presence': A project for the cultural landscape" *Environment and Planning D*, 24(4): 537–554.

Rozwadowski, H. (1996) "Small world: Forging a scientific maritime culture for oceanography" *Isis*, 87(3): 409–429.

Rozwadowski, H. (2003) "Salty dogs and 'Philosophers': A saga of seafaring scientists and sailors" *Wrack Lines*, 3(2): 1–5.

Rozwadowski, H. (2005) *Fathoming the Ocean: The Discovery and Exploration of the Deep Sea*. London: Harvard University Press.

Rudel, C. (2002) *Les Açores: un archipel au coeur de l'Atlantique*. Paris, France: Karthala.

Ryan, J. (1997) *Picturing Empire: Photography and the Visualization of the British Empire*. London: Reaktion.

Ryan, J. (2006) " 'Our home on the ocean': Lady Brassey and the voyages of the *Sunbeam*, 1847–1887" *Journal of Historical Geography*, 32: 579–614.

Ryan, J. and Schwartz, J. M. (Eds.) (2003) *Picturing Place: Photography and the Geographical Imagination*. London: I. B. Tauris.

Ryan, M. (1999) *Literary Theory: A Practical Introduction*. London: Blackwell.

Sacks, D. (1991) *The Widening Gate: Bristol and the Atlantic Economy, 1450–1700*. Berkeley, CA: University of California Press.

Said, E. (1994) *Culture and Imperialism*. London: Vintage.

Said, E. (2003) *Orientalism*. London: Penguin.

Salmond, A. (1991) *Two Worlds: First Meetings between Maori and Europeans, 1642–1772*. Honolulu, HI: University of Hawai'i Press.

Saussure, F. de (1910–1911/1998) "Course in General Linguistics" in Rivkin and Ryan (Eds.) (1998): 76–90.

Schein, R. (1997) "The place of landscape: A conceptual framework for interpreting an American scene" *Annals of the Association of American Geographers*, 87(4): 660–680.

Schrope, M. (2006) "The real sea change" *Nature*, 443: 622–624.

Secord, J. (2004) "Knowledge in transit" *Isis*, 95: 654–672.

Shapin, S. (1995) "Cordelia's love: Credibility and the social studies of science" *Perspectives on Science*, 3(3): 255–275.

Sobel, D. (2007) *Longitude: The True Story of a Lone Genius Who Solved the Greatest Scientific Problem of His Time*. London: Harper Perennial.

Sorlin, S. (2000) "Ordering the world for Europe: Science as intelligence and information as seen from the Northern Periphery" *Osiris*, 15 (2nd Series): 51–69.

Sorrenson, R. (1996) "The ship as a scientific instrument in the eighteenth century" *Osiris*, 11(4) (2nd Series): 221–236.

Steinberg, P. (1999a) "Navigating to multiple horizons: Toward a geography of ocean-space" *Professional Geographer*, 51(3): 366–375.

Steinberg, P. (1999b) "Lines of division, lines of connection: Stewardship in the world ocean" *Geographical Review*, 89(2): 254–264.

Steinberg, P. (2001) *The Social Construction of the Ocean*. Cambridge: Cambridge University Press.

Steinberg, P. (2009) "Sovereignty, territory, and the mapping of mobility: A view from the outside" *Annals of the Association of American Geographers*, 99(3): 467–495.

Stow, D. A. V. (2001) "Silent, Strong and Deep: The Mystery of how Basins Fill" in Deacon, Rice and Summerhayes (Eds.) (2001): 108–123.

Strommel, H. (1984) *Lost Islands: The Story of Islands That Have Vanished from Nautical Charts*. Vancouver, BC: University of British Columbia Press.

Summerhayes, C. (2001) "Introduction: The Exploration of the Sea Floor" in Deacon, Rice and Summerhayes (Eds.) (2001): 89–91.

Sverdrup, K., Duxbury, A. C. and Duxbury, A. (2003) *An Introduction to the World's Oceans*. 7th edition. International edition. New York, NY: McGraw-Hill.

Symmons, C. R. (1979) *Developments in International Law: Maritime Zones of Islands, Vol. 1*. Boston, MA: Brill.

Symmons, C. R. (1986) "The Rockall dispute deepens: An analysis of recent Danish and Icelandic actions" *The International and Comparative Law Quarterly*, 35(2): 344–373.

Taylor, D. (2000) "A Biogeographer's construction of Tropical Lands: A. R. Wallace, Biogeographical Method and the Malay Archipelago" *Singapore Journal of Tropical Geography*, 21(1): 63–75.

Thomson, C. Wyville, Murray, J., Nares, G. and Thomson, F. T. (1880–1895) *Report on the Scientific Results of the Voyage of H.M.S. Challenger during the Years 1873–76 Under the Command of Captain George S. Nares, R.N., F.R.S. and the Late Captain Frank Tourle Thomson*. Edinburgh: Royal Navy.

Thornton, J. K. (1998) *Africa and Africans in the Making of the Atlantic World, 1400–1800*. 2nd edition. Cambridge: Cambridge University Press.

Thrower, N. (Ed.) (1981) *The Three Voyages of Edmond Halley in the Paramore*. 2 Vols. London: The Hakluyt Society.

Thrower, N. (1999) *Maps and Civilization: Cartography and Culture and Society*. Chicago, IL: University of Chicago Press.

Trujillo, A. and Thurman, H. (2005) *Essentials of Oceanography*. 8th edition. London: Pearson Prentice Hall.

USC Libraries (2009) "Digital Collections" Web document: http://digarc.usc.edu. Last accessed 27/04/2009.

U.S. Geological Survey (USGS) (2014) "Understanding Plate Motions" Web document: http://pubs.usgs.gov/gip/dynamic/understanding.html. Last accessed 20/01/2016.

Vallega, A. (2001) "Ocean governance in post-modern society—a geographical perspective" *Marine Policy*, 25(6): 399–414.

Verne, J. (1990) *Vingt mille lieues sous les mers*. Paris, France: livre de poche.

von Humboldt, A. (1995) *Personal Narrative of a Journey to the Equinoctial Regions of the New Continent*. London: Penguin.

Walcott, D. (1986) *Collected Poems, 1948–1984*. New York, NY: Farrar, Staus and Giroux.

Wegener, A. (1966) *The Origin of Continents*. 4th edition. John Biram (Ed.). London: Methuen.

Whelan, K. (2004) "The Green Atlantic: Radical Reciprocities between Ireland and America in the Long Eighteenth Century" in Wilson (Ed.) (2004): 216–238.

Wijffels, S., Bryden, H., Schmitt, R. and Stigebrandt, A. (1992) "Transport of freshwater by the oceans" *Journal of Physical Oceanography*, 22(2): 155–162.

Williams, G. (2002). "Captain Cook, Explorer, Navigator and Pioneer" Web document from BBC History (2009).

Wilson, K. (Ed.) (2004) *A New Imperial History: Culture, Identity, and Modernity in Britain and the Empire, 1660–1840*. Cambridge: Cambridge University Press.

Withers, C. (1995) "Geography, natural history and the eighteenth-century Enlightenment: Putting the world in place" *History Workshop Journal*, 39: 136–163.

Withers, C. (1999) "Reporting, mapping, trusting: Making geographical knowledge in the late Seventeenth century" *Isis*, 90(3): 497–521.

Withers, C. (2006a) "Eighteenth-century geography: Texts, practices, sites" *Progress in Human Geography*, 30(6): 711–729.

Withers, C. (2006b) "Science at sea: Charting the Gulf Stream in the late Enlightenment" *Interdisciplinary Science Reviews*, 31(1): 58–76.

Withers, C. (2007) *Placing the Enlightenment: Thinking Geographically about the Age of Reason*. London: Chicago University Press.

World Ocean Observatory (WOO) (2016) "History of the World Ocean Observatory" Web document: http://www.worldoceanobservatory.org/content/history-world-ocean-observatory. Last accessed 20/01/2016.

Zerubavel, E. (1982) "The standardization of time: A sociohistorical perspective" *The American Journal of Sociology*, 88(1): 1–23.

Zerubavel, E. (2003) *Terra Cognita: The Mental Discovery of America*. London: Transaction Publishers.

Index

Page numbers in italics indicate figures.